LEVERAGING TRUST:
A Force Multiplier for Today

IN HARM'S WAY

Leveraging Trust:
A Force Multiplier for Today

Edited by
Lieutenant-Colonel Jeff Stouffer
and
Craig Leslie Mantle

CANADIAN DEFENCE ACADEMY PRESS

Canadian Defence Academy Press
PO Box 17000 Stn Forces
Kingston, Ontario K7K 7B4

Produced for the Canadian Defence Academy Press
by 17 Wing Winnipeg Publishing Office.
WPO30342

Cover Photo: 1RCR Soldier in the Arghendab during OP MEDUSA - Combat Camera, Sgt Lou Penney

Library and Archives Canada Cataloguing in Publication

Leveraging trust : a force multiplier for today / edited by Jeff Stouffer and Craig Leslie Mantle.

(In harm's way)
Issued by: Canadian Defence Academy.
Produced for the Canadian Defence Academy Press by 17 Wing Winnipeg Publishing Office.
Includes bibliographical references and index.
ISBN 978-0-662-48113-3 (bound) -- ISBN 978-0-662-48114-0 (pbk.)
Cat. no.: D4-4/5-1-2008E (bound) -- Cat. no.: D4-4/5-2-2008E (pbk.)

1. Canada--Armed Forces--Foreign countries. 2. Canada--Armed Forces--History--20th century. 3. Canada--Armed Forces--Officers. 4. Military art and science--Canada. 5. Operational art (Military science). 6. Command of troops. 7. Trust. I. Mantle, Craig Leslie, 1977- II. Stouffer, Jeff III. Canadian Defence Academy IV. Canada. Canadian Armed Forces. Wing, 17 V. Series.

U163 L38 2008 355.30971 C2008-980243-8

Printed in Canada.

3 5 7 9 10 8 6 4 2

ACKNOWLEDGEMENTS

The success of any project of this magnitude and scope is, with absolute certainty, the direct result of the individual contributions of many. Rarely in such pursuits is anything not a team effort. To this end, we would like to thank each of the contributing authors for their dedication and professionalism, not to mention the sacrifice of their personal time, in helping to bring this project to its fullest fruition. Without their valuable insights and desire to help others learn, this book would truly not have been possible. On behalf of the Canadian Defence Academy in general and the Canadian Forces Leadership Institute in particular, we would like to extend our sincere thanks to Master Corporal Shawn Raines who, although retired, freely offered his experiences without any expectation of reward, expending considerable time and energy in the process.

Greatly appreciated are the contributions of various individuals from Defence Research and Development Canada, specifically Megan Thompson, Barbara Adams and Jessica Sartori. Their support, encouragement and prior research on trust materially advanced this initiative and added much to the final product.

Of course, the success of any publication is also predicated on the degree of professional technical support that is received. As a result, we would like to express our heartfelt gratitude to Mélanie Denis of the Canadian Forces Leadership Institute (CFLI) for generously providing the necessary production expertise and savvy to keep this project (and us!) well on track. Her keen photographic eye, as well as her expertise *en français*, ensured that a graphically pleasing volume ultimately emerged. We also thank Greg Moore of CFLI, who came late to the project, but who aided us immensely in seeing this volume through the final stages of completion. As always, 17 Wing Publishing Office, especially Captain Phil Dawes, Evelyn Falk and Michael Bodnar, must be credited for their superb efforts in making this volume a reality, essentially morphing a "bundle" of papers into a finished and highly professional product.

TABLE OF CONTENTS

FOREWORD

I am extremely pleased to introduce *Leveraging Trust: A Force Multiplier for Today*, the fifth volume in the *In Harm's Way* series and another contribution to the broader Strategic Leadership Writing Project. Created by the Canadian Forces Leadership Institute, this latter collection seeks to create a uniquely Canadian body of leadership knowledge, to stimulate thought and debate on a variety of leadership topics, to aid in professional military development, and to increase general public awareness of some of the unique challenges faced by Canadian Forces personnel. In essence, this volume, like those that have preceded it, captures the experiences of serving members for the purpose of providing sage advice and guidance for all, particularly for those about to be placed in harm's way.

Within the pages of this volume, a diverse group of authors from a broad spectrum of rank levels and combat and support roles have provided their thoughts on the importance of trust, outlined the steps that they took to build and sustain trust over time, and described their actions taken in response to those occasions on which trust had been violated. I am convinced that their accounts will not only foster a better appreciation of the complexities and profound implications of trust, but will also provide timely and relevant examples of how trust is developed, maintained and repaired if need be. As will be seen, trust has a tremendous impact on desired individual and team outcomes; it is, after all, one of the most important force multipliers that we have at our disposal. Readers will hopefully come to a more complete appreciation of the need for leaders to understand the implications of both the presence and absence of trust in their daily relationships, especially in dangerous and demanding operations.

I commend this useful book to everyone regardless of rank, for without trust between leader and follower, success in today's complex and fluid security environment will be difficult, if not impossible, to achieve. The insights offered herein will help leaders at all levels to better appreciate how their actions, in word and in deed, can enhance or reduce trust and how they can better harness and capitalize on the power of this important and essential factor.

Major-General Daniel Gosselin
Commander, Canadian Defence Academy

INTRODUCTION

Lieutenant-Colonel Jeff Stouffer

Trust is a key enabler of success. It has often been described as the foundation upon which effective teams are built and "a good that organizations can't get enough of."[1] As Patrick Lencioni, author of the *New York Times* bestseller *The Five Dysfunctions of a Team*, asserts, "no quality or characteristic is more important than trust."[2] Indeed, researchers at Defence Research and Development Canada (DRDC)[3] argue that trust: a) enhances morale and cohesion; b) promotes open communications; c) promotes job satisfaction and organizational commitment; d) helps to induce problem solving that is productive and creative; e) has a positive effect on team performance; and, f) is related to numerous desired organizational citizenship and work behaviours such as conscientiousness, sportsmanship, civic virtue, courtesy, organizational commitment and helping others before oneself. Trust has also been shown to decrease defensive monitoring, the need for formalized controls, adversarial tactics and conflict. Trust therefore represents a critical part of an organization's social capital (or the psychological and behavioural contract between its members and the institution), and as such, "represents an important human dimension of military effectiveness."[4] Without debate, if properly harnessed and nurtured, trust will contribute to valued and necessary organizational goals. If ignored or taken for granted, however, trust can undermine individual, team and organizational effectiveness, as well as member well-being. Clearly, trust is an important attribute that must be created from the outset, sustained over the long term and rebuilt if for some reason it is destroyed. Those who fail in these regards compromise their chances of success.

All too often, there is a tendency "to mention trust in passing, to allude to it as a fundamental ingredient or lubricant, an unavoidable dimension of social interaction, only to move on to deal with less intractable matters."[5] This statement applies to many organizations, the Canadian Forces (CF) likely being no exception. As an example, the word "trust" is sprinkled liberally throughout CF literature with limited explanation or attention afforded to its antecedents, implications and the actual manner by which it operates. Are we so familiar with this construct that it goes largely ignored? Is it assumed that trust is something that just happens naturally and therefore requires little or no management? Rhetorical questions certainly, but important ones nonetheless.

Complicating this lack of professional attention is the reality that, like leadership, trust is a complex and elusive construct. In fact, there is no agreement within the expansive research literature on a single definition of trust. The one that is most widely accepted posits that "trust is a psychological state comprising the intention to accept vulnerability based upon positive expectations of the intentions or behaviour of another."[6] Francis Fukuyama, a renowned trust expert, adds that trust "is the expectation that arises within a community of regular, honest, and cooperative behaviour, based on commonly shared norms, on the part of other members of that community."[7] In this sense, leaders must acknowledge that trust is not limited to interactions between individuals and must be understood in terms of the general community. Without doubt, however, the need to trust others is strongest and most meaningful when something critical is at stake and there is a cost associated with failing to trust (i.e., the task requires the input of others to complete). As a concept, therefore, trust is most critical in situations that involve risk, vulnerability and the need to be interdependent with others.[8] With absolute certainty, current military operations in the "new security environment" involve considerable risk and vulnerability, as well as the need to depend on subordinates, peers and superiors to achieve mission success. Trust, therefore, is of paramount importance, not only in operations, but in garrison as well.

FACTORS THAT PROMOTE TRUST

To fully appreciate and harness trust, leaders must be cognizant of the many diverse factors associated with its development. Those factors found to influence judgments of trustworthiness are typically divided into three broad categories and include: a) qualities of the trustor (the person doing the trusting); b) qualities of the trustee (the person being trusted); and, c) qualities of the interaction.[9]

People differ in their propensity to trust and typically fall somewhere on a continuum ranging from being unwilling to trust at all to exhibiting blind trust. Research has demonstrated that one's predisposition to trust is initially evoked in new situations, but as time progresses, the need to rely on this tendency becomes less significant as relationships develop.[10] Previous trust experiences/histories that include both successes and violations of trust also influence the willingness to resist or cultivate future trust relationships (i.e., successive positive trust experiences reward and encourage trusting behaviour). With certainty, however, the qualities of the leader will in time influence subordinate propensity to trust, alter trust histories and have a positive or negative effect on ongoing trust experiences.

Qualities of the trustee (i.e., leader) commonly argued to influence trust include: a) competence; b) integrity; c) benevolence; and, d) predictability.[11] Competence is the belief that the trustee has the ability to complete a task or specific job (i.e., trust in another's abilities). Integrity involves consistency between a trustee's words and actions, and includes being fair, seeking subordinate input, and adhering to honest and open communication. Integrity "flows from values-based ethical behaviour."[12] "To have integrity," therefore, "is to have unconditional and steadfast commitment to a principled approach to meeting your obligations while being responsible and accountable for your actions."[13] Benevolence refers to showing a genuine concern for the trustor. For example, in describing the dynamics between leaders and their followers, clinical psychologist Henry Cloud argues that "trust is built when leaders have a genuine interest in knowing them, knowing about them, and having what we know matter."[14] Predictability simply refers to doing what one is expected to do (i.e., consistency in behaviour). Although competence, integrity, benevolence and predictability are distinct characteristics, it is important to acknowledge that trust can develop with varying degrees of each of these factors.

The way in which leaders and their followers interact will also influence the building of trust. Commonly noted influences include: a) communication styles; b) similarity; and, c) shared goals.[15] Open communication styles that facilitate the sharing of information, will help individuals find common ground and shared goals. As one sees others as being similar, the assumption is that they will behave in a similar manner, thus enabling a more accurate prediction of subsequent behaviour.

HOW LEADERS CAN PROMOTE TRUST

As Robert Levering, co-author of *The 100 Best Companies to Work for in America*, asserts, "A great place to work is one where you trust the people you work for."[16] Military leaders at all levels are in the obvious and unique position to act as the catalysts in promoting, developing and maintaining trust. As U.S. Army Lieutenant-General (Ret'd) Walter F. Ulmer asserts, "Competent military leaders develop trust."[17] Trust does not come with rank or appointment; it must be earned and is easily forfeit if one's conduct is found wanting. Being highly visible, leader actions set the tone for either the development or deterioration of trust. This reality makes it even more critical that leaders understand how their actions, in both deed and in word, can facilitate or debilitate the development of trust. Knowledge of the factors that contribute to the development of trust (as outlined above) can be effectively employed with moderate effort and the simple application of existing espoused CF leadership practices. The benefits, to be sure, will far outweigh the costs. Several means through which leaders can build trust include:

- demonstrating direct and tangible evidence of high levels of proficiency and professional competence in the performance of core functions and taking advantage of opportunities to enhance their professional expertise and competence;

- exercising good judgment in decisions that affect others and not exposing people to unnecessary risks including ensuring that taskings are consistent with individual or team abilities;

- showing trust and confidence in followers by giving them additional authority and involving them in decisions where circumstances allow;

- demonstrating concern for the well-being of followers, representing their interests and ensuring they are supported and taken care of by the organization;

- showing consideration and respect for others, being polite, friendly, approachable, as well as treating followers fairly and consistently (without favour or discrimination);

- being professional in bearing and conduct. Trusted leaders set high standards for themselves and demonstrate accountability. They also exemplify the best values and practices of the CF;

- maintaining honest and open communications. Trusted leaders communicate their intent, ensure it is understood, and strive to achieve transparency in communications to mitigate the debilitating effects that gossip and rumour can evoke;

- leading by example, sharing risks and hardships and refusing to accept or take special privileges;

- keeping their word and being counted upon to honour their obligations. Trusted leaders do what they say they are going to do. This includes applying rules in a fair and consistent manner; and,

- promoting commonalities that people share (values, goals, attitudes) and stressing tolerance for dissimilarities (i.e., understanding and leveraging factors associated with the development of both person- and category-based trust).[18]

CREATING AND SUSTAINING TRUST

Establishing trust is certainly possible, and for the most part, achievable through the application of established CF leadership practices. Research conducted by DRDC-Toronto offers several guidelines to support the development and maintenance of trust:[19]

- build trust as early as possible;

- promote professional development (for leaders and subordinates);

- ensure new members of teams have opportunities to display their competence and trustworthiness and assist them in identifying with the values and goals of the team;

- utilize open and transparent communications;

- identify trust as a central value and critical team objective;

- create a strong and positive value base;

- because trust takes time to develop, minimize turnover as much as possible in military teams. "The contemporary operating environment often throws soldiers into situations where they must quickly establish working relationships with complete strangers;"[20]

- demonstrate strong and positive leadership and maintain direct and personal contact with subordinates;

- promote a strong common identity (i.e., member of the CF);

- understand the antecedents of trust;
- ensure trust-eroding behaviour is eliminated (i.e., information hoarding, backstabbing, harassment, etc.); and,
- exercise vigilance to recognize warning signs of low or deteriorating trust (e.g., defensive monitoring, poor communications, etc.).

TRUST VIOLATIONS

Regardless of adherence to the above espoused actions or the best intentions of leaders, trust violations, whether real or perceived, will undoubtedly occur and represent a formidable leadership challenge. Violations of trust are often predicated on: rule violations; honour violations (shirking responsibilities, breaking promises, lying, stealing ideas, and secret/confidence disclosures); abusive authority; and, damaging another person's identity through public criticism, false or unfair accusations, and insults.[21]

SUMMARY

Awareness of the supporting trust literature as outlined above permits a more comprehensive understanding of how the actions and accounts reported in each of the following chapters, whether always intentional or at times unintentional, facilitated the building and maintenance of trust. Of marked interest is the striking congruence between trust building behaviours as reported in this volume and both the extant trust literature and current CF leadership principles. The parallels are indeed striking, and to a degree, comforting. To be sure, understanding and practicing these principles will set the stage for the development of trust within the CF in general and in small teams in particular. When present, trust facilitates mission success; its absence, in contrast, jeopardizes the attainment of specific goals.

It is fully expected that most leaders understand, to varying degrees of completeness, the implications and advantages of establishing trust. Some may argue, however, that this topic actually requires very little discussion and that what is presented in the pages that follow simply represents common sense and the application of time-proven leadership principles. If this is so, one must ask why CF surveys demonstrate less than favourable results when it comes to trust in leaders and the institution as a whole.[22]

If these principles are so well understood and easily practiced, why, then, is trust repeatedly identified as an attribute that is lacking? It is hoped that this present effort, combined with the ambitious program of research exploring trust in military operations currently being undertaken by DRDC, will serve to make the issue of trust more visible. As a force multiplier, it must receive due and proper attention if its full benefits are to be realized and used to advantage.

ENDNOTES

1 Tom L. Roberts, Paul D. Sweeney, Dean McFarlin and Paul H. Cheney, "Assessing Trust among IS Personnel: A View of General Trust, Trust of Management and Inter-Organizational Trust," *Proceedings of the 37th Hawaii International Conference on System Sciences*, 2004, 2.

2 Patrick Lencioni, *Overcoming the Five Dysfunctions of a Team* (San Francisco: Jossey-Bass, 2005), 13.

3 Barbara Adams, David Bryant and Robert Webb, *Trust in Teams Literature Review*, Department of National Defence [DND], 2001, DCIEM No. CR-2001-042.

4 Canada, *Leadership in the Canadian Forces: Conceptual Foundations* (Ottawa: DND, 2005), 74.

5 Diego Gambetta, "Can we Trust Trust?" in D. Gambetta, ed., *Trust: Making and Breaking Cooperative Relations* (New York: Basil Blackwell, 1998), 213-237.

6 Denise M. Rousseau, Sim B. Sitkin, Ronald S. Burt and Colin Camerer, "Not so Different After All: A Cross-Discipline View of Trust," *Academy of Management Review* 23, no. 3 (1998), 394.

7 Francis Fukuyama, *Trust: The Social Virtues and the Creation of Prosperity* (New York: Free Press Paperbacks, 1995), 26.

8 Adams, Bryant and Webb, *Trust in Teams Literature Review*, 16.

9 Ibid., 26.

10 Lynne C. D'Amico, "Examining Determinants of Managerial Trust: Evidence from a Laboratory Experiment," *National Public Management Research Conference*, Washington, DC, 10 October 2003, 2.

11 Adams, Bryant and Webb, *Trust in Teams Literature Review*, 28.

12 Christopher R. Kemp, "Trust – The Key to Leadership in Network Centric Environments," (Unpublished Thesis, United States Army War College, Carlisle Barracks, PA, 2003), 16.

13 Canada, *Duty with Honour: The Profession of Arms in Canada* (Ottawa: DND, 2003), 31.

14 Henry Cloud, *Integrity* (New York: Collins, 2006), 56.

15 Adams, Bryant and Webb, *Trust in Teams Literature Review*, 30-31.

16 Robert Levering and Milton Moskowitz, *The 100 Best Companies to Work for in America* (New York: Currency/Doubleday, 1993).

17 Lieutenant-General (Ret'd) Walter F. Ulmer, as quoted in *Leadership in the Canadian Forces: Conceptual Foundations*, vii.

18 Canada, *Leadership in the Canadian Forces: Leading People* (Ottawa: DND, 2007), 70-71.

19 Adams, Bryant and Webb, *Trust in Teams Literature Review*, 85-86.

20 Colonel Christopher R. Paparone, "The Nature of Soldierly Trust," *Military Review* (November-December 2002), 45.

21 Robert J. Bies and Thomas M. Tripp, "Beyond Distrust: 'Getting Even' and the Need for Revenge", in R. Kramer and T. Tyler, eds., *Trust in Organizations* (Thousand Oaks: CA: Sage, 1996), 249.

22 Unpublished results from the 2006 Canadian Forces Retention Survey indicate that 67% of the sample somewhat or strongly agreed with the statement "if given the opportunity, the CF would take advantage of me." The survey also showed that 47% of the sample somewhat to strongly disagreed with the statement "I trust senior leaders to make the right decisions for CF members." Equally disconcerting, 36% somewhat to strongly disagreed with the statement "I trust unit leaders to make the right decisions for the unit."

CHAPTER 1

TRUST IN THE SANDSTORMS OF AFGHANISTAN: OPERATION ATHENA ROTATION 3

Lieutenant-Colonel Jim Kile

Afghanistan became a household name in Canada soon after 9/11. Through one of the most shocking terrorist attacks on a Western nation to date, Osama bin Laden brought war to the continental shores of the United States (US). Nineteen terrorists from Saudi Arabia, Egypt and other Arab states, working in small groups, hijacked four commercial airliners and unleashed their unsuspecting power on previously identified symbols of American pride. This audacious and diabolical plan resulted in the deaths of almost 3,000 people and culminated in the total destruction of the twin towers of the World Trade Center in New York City; the Pentagon was also heavily damaged. The terrorists that conducted this attack were affiliated with the Al-Qaeda (Arabic for "the base") network, a radical Islamic group led by Saudi exile Osama bin Laden. He and his *jihadists* were dedicated to waging a holy war against the US and its allies, and to this day remain just as committed. Al-Qaeda was able to plan and train with impunity for these attacks in Taliban-controlled Afghanistan in a region bordering with Pakistan.

Seeing their country under attack, Americans rallied around their leaders with a sense of patriotism and outrage. The US government immediately demanded that Afghanistan turn over Osama bin Laden. Taliban leaders refused and, in October 2001, the US, enlisting the help of the Northern Alliance and using effective air power, particularly laser guided munitions, quickly removed the Taliban from power. This US-led mission was heralded by the entry of Special Operations personnel into Afghanistan on 10 October 2001.[1]

Canada's initial contribution to this 21st century war on terror was termed Operation Apollo.[2] Soldiers predominately of the Princess Patricia's Canadian Light Infantry (PPCLI) from Winnipeg (later Shilo) quickly established a base at Kandahar Airfield. Events moved rapidly as Afghan nationals representing the Afghan people at the Bonn Summit in 2001 voted to establish the Afghanistan Transitional Authority (ATA) with

Hamid Karzai, a former Pashtun Mujahedeen commander, as transitional leader. In 2002, he was elected the Transitional Head of State and remains in power today. Once realizing that its nominal hold on Afghanistan was precarious and that the enemy, although defeated, was everywhere, the ATA immediately turned to the international community for assistance. The United Nations Security Council consequently mandated the International Security Assistance Force (ISAF), representing the North Atlantic Treaty Organization (NATO), to support the ATA by improving security and assisting in rebuilding the country.

ISAF continues to operate in Afghanistan; it is not a UN operation. Afghanistan is not, nor has it ever been, a traditional peacekeeping mission for there is no peace to keep and negotiation with an enemy such as the Taliban or Al-Qaeda is not an option for many. ISAF is often described as a "coalition of the willing" deployed under the authority of the UN Security Council. At its outset in August 2003, the "willing," or rather those countries that had contributed troops, consisted of 36 nations including Canada. The primary focus of ISAF was the security of Afghanistan's capital city and the seat of government power, Kabul. Concurrently, the Americans were fighting in the south, in Kandahar and Helmand provinces, as part of Operation Enduring Freedom.

Operation Athena was Canada's contribution to ISAF's initial mandate in the Afghan capital. Canada sent troops immediately to an area in the southwestern corner of the city. There, the military established a base and named it Camp Julien. The Government of Canada's overarching goal was, and continues to be, preventing Afghanistan from relapsing into a failed state that provides a safe haven for terrorists and terrorist organizations.

Keys to early success in the Kabul area for ISAF's 8,000 troops included: the creation of freedom of movement in Kabul and the surrounding area; the reparation, protection and maintenance of Kabul International Airport; and, assisting in overall reconstruction and in the rebuilding of the national armed forces of Afghanistan in order to ensure national security and stability. The identified end-state was the creation and continued existence of a viable democracy in which Afghan citizens could rely on the government and the rule of law for safety.

I was asked to become part of Operation Athena, Rotation (Roto) 3, in September 2004. Within two weeks, I entered Afghanistan as part of a tactical reconnaissance team. It was a country that I knew precious little about, but would become all too familiar with in very short order. In the fall of 2004, when I received my warning order, Canada's focus was Kabul. The Americans were in Kandahar where the fighting was far heavier. That being said, it was evident from the beginning that Afghanistan was going to be a dangerous and volatile place where confidence in my abilities as a leader within my company and trust in my team would be essential elements of our success.

The medical company (i.e., Health Services Support (HSS Coy)) for Roto 3 stood up on 12 November 2004. This day heralded an intense period as 53 individuals came together for the first time in order to prepare for what lay ahead. It is noteworthy that back in 2004, while Afghanistan was the Canadian Forces' (CF) primary mandate, there was, I would suggest from an individual perspective, a lack of familiarity and understanding as to what could be expected in theatre. From a medical perspective, casualty rates were difficult to predict. CF health services planners, taking into account other allied medical capabilities in the Kabul area, decided to send surgical capability to Afghanistan.

SETTING THE CONTEXT: AN INTRODUCTION TO TRUST

The word "trust" comes from an old 12th century word "traust" meaning confidence. People intuitively understand this definition. Most appreciate that trust can be as fragile as crystal and easily shattered. Carefully managed, however, it can be as strong and enduring as diamond. In day-to-day activity when two or more people come together, trust can take many forms and is often very elusive. It is a word that has been misused and at times overused when describing group dynamics. At the heart of any highly effective group is trust. Trust has been described as the highest form of motivation;[3] that is to say, a group without trust is a group that is dysfunctional.

Groups come together in many ways for many reasons. HSS Coy was a group that came together from military bases across Canada to provide forward surgical and medical capability to Task Force Kabul (TFK) Roto 3 from February to August 2005. This capability, usually referred to in medical phraseology as "Role 2 plus," was comprised of 53 personnel representing at least 15 different HSS trades. Given its over-all mission of "preserving manpower," being "faithful in adversity," while hastening "to aid the soldier,"[4] it was imperative that members of this newly formed group quickly grow to trust one another. The new command team inherently understood this particular challenge. While not specifically articulating this notion of trust *per se*, the team acted quickly to begin this complicated process.

HSS Coy's specific mission for Roto 3 in Afghanistan was to provide robust medical and surgical capability to Canadian soldiers and other troop contributing nations located at Camp Julien. The camp, as mentioned, was strategically situated in the southwestern corner of Kabul. The total patient population was roughly 1,500. In addition, HSS Coy was to be an integral part of ISAF's mass casualty[5] plan in the Kabul area of responsibility (AOR). Further, as part of the TFK commander's outreach plan, HSS Coy was expected to provide a degree of humanitarian support to the local Afghan population. The members of HSS Coy clearly had to function well in their jobs, but more important, they had to function well as a team in order meet the immense demands.

To establish trust with such an unfamiliar and diversified group would be one of several goals of any commanding officer (CO), and for me, trust was absolutely vital and fundamental to the success of our team. Historically, this was not a new challenge for Canada's medical services. 38 Field Ambulance (38 Fd Amb), which possessed similar forward surgical capability as it prepared for deployment to Korea in May 1952, was "brought up to strength by having its Officers and Men posted in from across Canada, from various medical installations, hospitals, personnel depots, command and area detachments."[6] Given the complex nature of health services and the unique and potentially volatile nature of any mission to Afghanistan, it was evident that I had precious little time to waste as CO in forming the group into an effective and cohesive team capable of supporting each other and the mission. Establishing trust takes time and patience. It cannot preclude the necessity to train and develop people in the hopes of improving their competence.[7] Professionalism and high skill levels provide a more fertile ground in which trust can grow. Under these circumstances, the notion of establishing trust is neither overused nor misplaced. What is also certain is that as a commander, I needed to have an understanding of trust in the group context and some type of implementation plan.

For all teams, it is not technology, finances or strategy, but rather teamwork, which remains the ultimate competitive or synergistic advantage, both because it is so powerful and so rare.[8] For all its attention over the years, ranging from passages in Sun Tzu's classic *The Art of War*[9] to commentary by Colin Powell,[10] teamwork remains as elusive as it has ever been within most organizations, whether civilian or military. The CF in recent times has recognized this and now demonstrates a clear understanding of this concept.[11] Consistent in all teamwork and team dynamic literature is the idea that it is the presence of trust among team members that leads to synergy and at times a cascading effect resulting in maximum performance. Conversely, the lack of trust is seen as dysfunctional, and like a chain with one link broken, its absence will lead to the deterioration of teamwork.

In *The Art of War*, Sun Tzu appreciates the challenges faced by the "sage commanders" of Northern China some 2,300 years ago. Remarkably, trust is addressed and described as the "conventional patience," in that the commander is able to engender trust amongst his troops through genuineness and authenticity.[12] This appreciation of trust is no less applicable today. The notion of trust within modern groups can be considered in three ways. First, trust can be viewed as the willingness of team or group members to be seen as vulnerable within the group.[13] This could stem from a positive expectation (or belief) that another will not act opportunistically. Second, another possibility is that trust can be seen as forming a basis for predicting how an individual will act when faced with adversity. This is tied to a more basic understanding of trust in which a trustworthy person is perceived as making decisions or taking action based on confidence in and reliance on good qualities, especially fairness, truth, honour and ability. Further, the degree of predictability is seen as being directly proportional to the scale and extent of training and experience. Third, trust can be seen to operate on an intra-individual level in which a person within the group is often faced with placing confidence in one's own abilities in order to make a difficult decision. Another way to look at this aspect of trust is to consider the notion of "trusting" one's own intuition when little else is available in order to determine a risky but necessary course of action.[14]

The aim of the remainder of this chapter is to expand on the ideas described above by providing some personal insights, anecdotes and examples related to my dual role as the CO of HSS Coy and the senior medical advisor to the TFK commander. In this way, I will address each of the three elements of trust.

ELEMENT #1: TRUST AND VULNERABILITY WITHIN THE GROUP

Issues of trust arise in groups when people are dependent on others, or have to work with others, to achieve important outcomes.[15] This idea is universal and is no less relevant in industry, the combat arms, sports teams, or in my case, health services. Looking at the first type of trust from a positive perspective, I appreciated that if members of a team truly trusted one another, they would engage in unfiltered conflict around ideas. This would foster an environment of freethinking and brainstorming, two ingredients necessary for creativity. Ideally, this trust would lead to a greater commitment to decisions and plans of action. Further, individuals within the group would hold one another accountable for delivering against those plans with a focus on the achievement of collective results. At face value and historically, establishing this element of trust would clearly be a daunting task, particularly when HSS Coy for Roto 3 consisted of 54% augmentees. In this case, 2 Field Ambulance (2 Fd Amb) in Petawawa was the main contributing unit, accounting for 46% of personnel. This situation was not new to medical units preparing for Afghanistan. I suspect that the CO of 38 Fd Amb (referred to earlier) who gathered with his troops in Borden for training prior to their deployment to Korea in 1952 faced similar challenges.

My first priority was to establish a strong relationship with key members of my command team and to convey to them my vision. I was fortunate, in that my deputy commanding officer (DCO) was a former colleague with a solid reputation for professionalism, discernment and fairness. In terms of the company sergeant-major (CSM), I was left with the final word on who would join the company in this capacity. Ultimately, I was able to secure the services of a CSM known for his straight shooting, honesty and gregariousness. As mentioned, 54% of my newly formed company were augmentees. With forging an alliance with these two experienced men as one of my priorities, I began the process of instituting a culture of trust within the company. Similarly, I needed to identify my key sub-commanders, and together as a command team, ensure that trust was a universal team norm.

The key to establishing the first element of trust is to engender a culture or belief within the unit that the goal is not perfection but rather the achievement of excellence. In moving towards the goal, it is important that members within the group feel that mistakes and errors are "OK."[16] This notion promotes cooperation and decreases the likelihood of risk adverse or mitigating behaviour. My approach to this challenge was to communicate my philosophy from the beginning and, as a follow-up, to avoid micromanagement at any level and at all costs. General George Patton once said, "Tell your people what you want and not how to do it, and they will always surprise you with their ingenuity."[17] Below is a list of specific steps that I took in order to begin the process of ensuring that my HSS team would operate with an underlying attitude of trust and support:

- Within 48 hours of company stand-up, I interviewed all officers and key non-commissioned members (NCMs). I listened to their backgrounds and experiences and attempted to get a sense of their current capabilities and confidence level. In addition, I spoke to them about my command and leadership philosophy.

- During the first week of pre-deployment activity, members of the newly formed company attended a "meet and greet" to provide an initial opportunity to get to know each other. In this less formal environment, I first spoke on my expectations in terms of having a group norm of trust. I expressed a simple message, that I encouraged and expected open and honest expression of ideas.

- Throughout the pre-deployment training period, all members of the company gathered each morning for a group coffee and an opportunity to review the training for the day. These occasions along with other group meetings afforded me the opportunity to speak regularly to reinforce the norms I expected the HSS team to possess.

- In order to provide an increased appreciation for the known challenges that lay ahead and to decrease anxiety related to unexpected situations, I provided a detailed and in-depth summary of Afghanistan history, current problems, medical intelligence and the nature of our medical deployment. Members were exposed to hundreds of related digital photos accompanied by relevant commentary.

- Furthermore, on the first day, I addressed the entire company on issues related to the chain of command and the professional technical network. They were provided with my perspective and where they were going in terms of the mission and whom they were going to support. Highlights of my presentation included:

- "Treat all those you encounter as if some day you could be subordinate to them."

- "No one is smart enough to design a complex organization ahead of time. This is a living process in which adjustments will have to be made as the new organization moves through time. There must be central tendencies with significant flexibility/elasticity."

- "Keep a sense of humour."

- "Strive for excellence not perfection…for perfection is God's work."

- "Get fit and stay fit."

This list is certainly not all-inclusive, but in my estimation, it provided sufficient insight to members of my company in terms of where I was coming from and which direction I expected the company to take. The initial aim was to create and maintain clear and appropriate channels of communication, increase understanding of the mission, and enhance comfort levels with each other. Lastly, this represented my first attempt among several to encourage the building of trust within the team through a general demonstration of my own vulnerability early in the process. Added to this was an effort to create an environment that did not punish such vulnerability. I hoped to achieve this goal by employing an adult education approach and treating each mistake as a constructive learning opportunity.[18] Using this methodology, every teaching moment is an opportunity to provide constructive and positive feedback to individuals and the group.

It is important to appreciate that fostering trust is simply the first step in a series of steps that will increase the likelihood of cohesive and effective team performance. It is the first step leading to success, providing that a leader recognizes the need to foster healthy conflict; encourage clarity and buy-in; create mechanisms at the lowest levels for accountability; and, ensure that members of the team remain focused on collective goals.[19]

In hindsight, it was apparent that we gelled quickly. After the validation period, the Task Force Commander, Colonel Walter Semianiw (now

Major-General) commented to the Director General of Health Services in an e-mail that he was impressed with the medical unit. As a company, I believed that we continued to build on this. By the end of the six-month tour, HSS had faced several interesting and at times difficult challenges, both minor and major in nature. Day-to-day sick parade (i.e., walk-in medical appointments) was busy. Medical staff would see a gambit of clinical cases ranging from scorpion bites to anaphylaxis. The hospital dealt with two mass casualty scenarios. The first incident was an infectious gastroenteritis outbreak with the likely culprit being a common virus. The second incident involved a rollover of an armoured personnel carrier carrying several passengers. The surgical team provided treatment for casualties from improvised explosive devices and land mines. The intensive care unit medical team successfully managed a civilian casualty that experienced multi-organ failure. Our mental health capacity was readily available and capable of managing the degree and type of cases encountered in Afghanistan. All the medical staff to varying degrees contributed to some form of humanitarian aid to the local population. In the end, we saw some adversity and met all challenges with success.

ELEMENT #2: TRUST AS A MEANS OF PREDICTING BEHAVIOUR

Predictability is very desirable. Individuals prefer and function best under predictable circumstances. People like "cognitive consistency"[20] in their lives. The leader of a group, and in particular a commanding officer, given his or her authority, has wide ranging opportunities to maximize predictability and consistency. This can truly be challenging, as armies by their nature will encounter the *fog of war*. Inconsistency on a daily basis within the work environment increases what has been described as an individual's "fret" level.[21] A chronically elevated fret level, combined with the threat level encountered in Afghanistan, is a recipe for poor group dynamics and related dysfunctions.

The CF's principles of leadership speak to this very situation: keeping your followers informed; achieving professional competence; ensuring your followers know your meaning and intent; and, making sound and timely decisions are only a few examples of these axioms.[22] These principles inherently relate to a commander's action. Action can be seen as a behaviour that is recognizable and tangible. An intangible component to all of this is when followers can comfortably predict the leader's behaviour based on a well-defined and universally positive set of values. This constitutes a more classic view of trust. Colin Powell relates to this when he asks, "Why would you follow somebody around a corner? The reason is trust."[23] To build trust, Powell writes, "a leader must demonstrate the key attributes and personal traits that are likely to build people's confidence in you."[24] He was referring specifically to competence, character, courage, loyalty, confidence, selflessness, sacrifice and empathy. *Leadership in the Canadian Forces: Conceptual Foundations* outlines similar characteristics, in addition to integrity, dependability and fairness.[25] In essence, leadership must model positive team norms at all times.[26]

An important overriding principle is that ethical behaviour cannot be compartmentalized and adherence to a value system cannot be based on convenience. A leader must consistently reflect, demonstrate and communicate those values that are in-line with military ethos and culture. Honour, integrity, loyalty and obedience cannot be seen as "punch lines."[27] They do matter. They must be internalized and then reflected daily in everything a commander does. For me, I found this particular form of instilling trust less daunting because it related to behaviour exclusively under my control. Realizing that the goal is excellence and not perfection, and allowing for some vulnerability and error in judgment, I did not feel pressured to "behave." Rather, it made difficult decisions less difficult because all my decisions had a basis in fairness and trust. *There can be no hidden agendas.*

Establishing predictability and consistency relates to all things that we do as CF personnel in order to prepare for operations. For HSS Coy, these included: meaningful individual and collective training; practical and effective physical training; well-defined and heavily rehearsed mass casualty scenarios; focused and relevant medical in-service instruction; and, a concise and clearly communicated commander's intent. Using the stage model of operational deployment as a backdrop, I often reminded the entire company of how much training had been done and always reflected successes to date. For example, by mid-tour, the list of training, taskings and accomplishments included: the provision of in-theatre combat life-saver training for the entire task force; a live video teleconference with the Ottawa hospital; multiple physical education workshops; training on NATO standard helicopter medical evacuation; hosting International Nursing Day in which the guest of honour was the senior nurse in the Afghanistan National Army; ongoing collective training;

the establishment of an effective daily routine; winning an international ball hockey tournament; the provision of high quality preventive medicine and vermin control methods; and, the provision of first-rate patient care to several injured ISAF and Canadian soldiers. As mentioned, I would present and add to this list quarterly while in theatre. Each point was graphically represented as bricks building the foundation of a strong edifice. I would also use this opportunity to remind the Company of the constructive group norms that helped us in our successes to that point.

ELEMENT #3: TRUSTING ONE'S INTUITION

As a medical officer and scientist, the majority of my decisions and actions are based on empirical data. Data can be seen simply as reliable information. There are times, however, when leaders are faced with an information gap or an absence of data to feed into the decision-making process. This is the grey area. Ambiguities, depending on their nature, can make anyone feel uncomfortable. *The Art of War* relates similar circumstances when it describes "chaos" as being the greatest challenge for the "sage commander." This book further emphasizes that, "although chaos is generally a difficult and uncomfortable time, it is also dynamic, a time of great openness and creativity."[28] In order to be successful, the "sage commander" must not be undermined by uncertainty. The passage implies that at times, an experienced leader must trust intuition. Colin Powell refers to this as a classic "gut check," in that the effective leader looks hard at the evidence that is being presented by the people under his command and assesses its authenticity based on context and available information.[29] However, in times of "chaos" or "ambiguity," where context and information are lacking and a decision is imminent, leaders may need to trust their gut.[30] Fortunately, relying wholly on instinct or intuition is a rare event. Ensuring reliable intuition, one that you would trust, develops with time. As a result, maturity and life lesson insights are paramount. From a systems perspective (i.e., the CF), it means appropriate time in rank. From a personal perspective, it means appropriate time spent in the learning mode. Overall, it means a reliance and adherence to an appropriate and well-defined value system.

In terms of Afghanistan and my particular command, all of this relates to what I brought to the table at that time. It reflects my previous training, experience and values. For me, this was my first command, but intuitively,

I knew the day the company stood-up that I could trust in my abilities providing I adhered to the combined values and norms of the CF and the vocation of medicine. It was my aim to bring everyone home safely, and ideally, with a sense of pride, accomplishment and belonging.

This notion of trusting intuition was tested during a cholera outbreak that occurred in Kabul during the spring of 2005. The city and surrounding area was facing an outbreak of cholera, which is a severe form of watery diarrhea. It can be debilitating and is often lethal to the very young and very old. NATO medical sources reported roughly 7,000 cases of sudden onset watery diarrhea in the civilian population within the Kabul region over a 30-day period in May and June of 2005. Fifty-one cases were tested and 35 were positive for cholera infection. The source of infection was the city's wells. The primary cause of the well contamination was the greater-than-average winter run-off from the surrounding mountains. The Afghanistan Ministry of Health reported six related deaths. Local non-governmental organizations (NGOs) were divided as to whether or not this was a typical seasonal outbreak or an epidemic. The Canadian press had picked up the story and I found myself speaking with reporters from the *Globe and Mail* and the *National Post*. During my interviews, I had suggested that in my view, the Afghanistan Government had moved quickly and appropriately to put measures in place to deal with the cholera outbreak. Indeed, my statement was intuitive, based on my own experiences and training, along with an understanding of the recent advancement in Afghanistan's over-all medical capabilities. The statement turned out to be accurate. I based the comment on the reality that cholera outbreaks were typical in this region and the Afghans were familiar with containing them. In addition, the Afghan Ministry of Health collaborated with NATO subject matter experts for advice on many health-related issues. In this case, this partnership was called the Joint Cholera Task Force. Moreover, in my view, the cholera outbreak was being sensationalized in the media by certain NGOs for political and financial gain. In the end, the task force was successful in dealing with the cholera outbreak, and during a wrap up meeting, a representative from the Ministry of Health informed me that a besieged but grateful health minister wished to thank me for my comments.

It is worth noting that a small, unrelated outbreak of sudden onset watery diarrhea occurred at Camp Julien and at two other ISAF bases in the Kabul area of operations concurrently with the Kabul cholera outbreak.

The contagion was later identified as a strain of Norwalk virus. The contamination likely originated from an infected soldier returning from Canada or Europe. Intuitively, I knew that people would invariably connect the two outbreaks leading to increased anxiety levels and perhaps a sense of panic. Intuitively, I felt that quick pre-emptive action was necessary. I consequently utilized my experienced preventative medicine technician and my infectious disease subject matter expert to begin an information campaign that was proactive rather than reactive. I would suggest that we, as a unit, maintained and continually confirmed the confidence of the Task Force in our abilities. Our actions attenuated any public worry. The few concerns coming from the Task Force over health matters indicated to me that we were successful in maintaining their overall trust and allying their fears.

SUMMARY

While this chapter focused on issues around trust and attempted to provide operational definitions with some specific examples, it is important to realize that trust and leadership are inter-related in a complex manner. In my view, the capable leader or "sage commander" moves comfortably from situation to situation never missing an opportunity, even at the lowest level, to instill confidence and trust. Trust is naturally thought of in terms of character, but it is equally important that trust is thought of in terms of competence.[31] It follows that training and hard work are essential. The degree to which I was successful in many ways is for others to determine, particularly by those who were under my command, along with my commander. I believe our performance as a HSS team was exceptional, but then one could argue that I am somewhat biased! The Human Dimensions of Operations Survey completed at the end of the tour demonstrated that the mean level of perceptions for HSS Coy was high across all dimensions which included: unit morale and cohesion; task cohesion; professional morale; military ethos; and, leadership skills. The medical mass casualty and evacuation plan was one of the most complex team related efforts. Our plan was adopted as the NATO standard for Kabul AOR. It reflected considerable teamwork and resulted from open and honest expression and contribution from all the key stakeholders, which included all 53 members of the unit. It is clear that trust played a major role in our success. Relationships of all kinds are built on and sustained by trust. While there is inherent risk in trusting others, greater risk is probable when not trusting others. As I look back, it would be an honour to command this group again.

ENDNOTES

1 G.C. Schroen, *First In: An Insider's Account of How the CIA Spearheaded the War on Terror in Afghanistan* (New York: Presidio Press, Ballatine Books, 2006).

2 Department of National Defence [DND] Backgrounder, Op APOLLO.

3 P. Lencioni, *Overcoming the Five Dysfunctions of a Team* (San Francisco: Jossey-Bass, 2005).

4 C. Lanoue, *Canadian Forces Medical Services: Introduction to its History and Heritage* (Ottawa: Director General Health Services, 2002).

5 A "mascal" or mass casualty situation is one in which an excessive disparity exists between the casualty load and the medical capabilities locally available for its conventional management. This disparity may involve any of the following factors alone or in combination: personnel, facilities, equipment, supplies, communications and evacuation means.

6 Bill Rawling, *The Myriad Challenges of Peace: Canadian Forces Medical Practitioners Since the Second World War* (Ottawa: Her Majesty the Queen in Right of Canada, represented by the Minister of Public Works and Government Services, Canadian Government Publishing, Communications Canada, 2004), 66.

7 Stephen R. Covey, *The 7 Habits of Highly Effective People* (New York: Free Press, 2004), 178.

8 Lencioni, *Five Dysfunctions*, 195.

9 Sun Tzu, *The Art of War* (Oxford: Oxford University Press, 1971).

10 O. Harari, *The Leadership Secrets of Colin Powell* (New York: McGraw Hill, 2002).

11 Canada, *Leadership in the Canadian Forces: Conceptual Foundations* (Ottawa: DND, 2005).

12 Sun Tzu, *The Art of War*, 86.

13 Lencioni, *Five Dysfunctions*, 196.

14 D. Rock, *Quiet Leadership – Six Steps to Transforming Performance at Work* (New York: HarperCollins, 2006).

15 Canada, *Conceptual Foundations*, 158.

16 For a discussion surrounding culture and the acceptability of error, see the chapter by Colonel Bernd Horn in this present volume.

17 A. Axelrod, *Patton: A Biography* (Basingstokes, Hampshire: Palgrave Macmillan, 2006), 205.

18. Again, see the chapter by Colonel Horn.

19 Lencioni, *Five Dysfunctions*, 196.

20 "Cognitive Consistency" is an older and outdated theoretical notion that describes an innate human tendency to seek out stimuli that are consistent with one's beliefs and attitudes and to censor or limit one's exposure to stimuli that are inconsistent with one's beliefs and attitudes. *Wikipedia*, 2007.

21 Physician Management Institute, Ontario Medical Association, 2002.

22 Canada, *Conceptual Foundations*, Annex A, The "Principles of Leadership" in a Complex World, 55. Such principles include: achieve professional competence; know your own strengths and limitations and pursue self-improvement; seek and accept responsibility; lead by example; make sure that your subordinates know your meaning and intent, and then lead them to the accomplishment of the mission; know your subordinates and promote their welfare; develop the leadership potential of your personnel; make sound and timely decisions; train your subordinates as a team and employ them up to their capabilities; and, keep your personnel informed of the mission, the changing situation, and the overall picture.

23 Harari, *Leadership Secrets*, 203.

24 Ibid.

25 Canada, *Conceptual Foundations*, 61.

26 D.A. Feldman, *The Manager's Pocket Guide to Workplace Coaching* (Amherst, Massachusetts: HRD Press, 2001).

27 Quotation taken from the motion picture, "A Few Good Men," directed by Rob Reiner, Castle Rock Entertainment, Columbia Pictures, and New Line Cinema, 1992.

28 Sun Tzu, *The Art of War*, 91.

29 Harari, *Leadership Secrets*, 49.

30 Rock, *Quiet Leadership*, 199.

31 Stephen R. Convey, *The Speed of Trust – The Thing that Changes Everything* (New York: Free Press, 2007), 30.

CHAPTER 2

LIAR, LIAR: A VIOLATION OF TRUST ON OPERATION STABLE

Master Warrant Officer Martin Colbert with Craig Leslie Mantle[1]

Between July 1996 and July 1997, Canada contributed military personnel to the United Nations Support Mission in Haiti (UNSMIH). Known as Operations Standard and Stable, the mission's overall mandate included maintaining a secure environment, assisting the government in professionalizing and training the Haitian National Police, and supporting the Special Representative of the UN Secretary-General in coordinating the activities of various UN agencies involved with institution-building, economic rehabilitation and national reconciliation.[2] Haiti's history over the past half-decade or so had been extremely turbulent, witnessing a *coup d'état*, the enforcement of economic sanctions by the UN against the Haitian state, presidential elections, and the attempt by various iterations of "forces," UN and otherwise, to restore peace and good government. UNSMIH was, in mid-1996, the latest attempt by the international community to support Haiti as it consolidated its fragile democracy and restored both its judicial system and economy that for years had been neglected and mismanaged.[3] To achieve these various and diverse ends, Canada committed the 3rd Battalion, Royal 22nd Regiment (3 R22ndR) and various aviation, engineering and security assets.[4]

Then a sergeant but now a master warrant officer, I deployed on Op Stable as a section commander in 3 R22ndR. This was my second tour, having already served in Cyprus with the Canadian Airborne Regiment in 1986. My section consisted of 12 soldiers, about 90 percent of whom were airborne qualified, and the bond between us all was exceeding tight.[5] As part of the overall UN mission, my soldiers and I provided personal security to the newly elected president, René Préval, for a period of three months. Prior to leaving Canada, however, my section had trained exceedingly hard with JTF-2, Canada's special operations forces, and with time we became highly proficient in counter-ambush tactics.[6]

Because of the importance of this tasking, my soldiers and I (who were constantly on 24-hour stand-by) stored all of our equipment, weapons and supplies in the same small room on the first floor of the Presidential Palace where we lived, thus enabling our rapid deployment. Given the absolute requirement to be quick to action, the potential for mistakes was somewhat higher than usual and accidents became more of a concern. Being highly trained, competent and professional, everyone understood the need to be as safe as possible, but it was not long before an error occurred that duly set in motion a chain of events that would lead to a gross violation of trust.

Being the senior sergeant, I was appointed the acting platoon warrant officer for three weeks when the regular one was on out-of-theatre leave. My second-in-command (2IC), a master corporal, duly took over the section in my absence and was thus made responsible for providing presidential security. Nothing seemed amiss at the moment and the leadership transition went smoothly. This particular soldier was highly trained, having earned a number of important airborne qualifications, was extremely competent and displayed both excellent drive and unequalled initiative. Many in the section, and even the larger platoon, characterized the master corporal as one of the best amongst them, a model soldier and an example for all. Rarely did I have to correct his ways or provide close supervision. Based on our past history and good working relationship, I naturally placed a good deal of trust and confidence in him and certainly had no reason to doubt that he would continue his exemplary performance in his new, yet temporary, position.

On receiving notification that the president required protection as he was venturing outside of the secure Palace where he resided, the counter-ambush team duly prepared itself as it had many times before. My 2IC rallied his subordinates, yet in the excitement of the moment, he accidentally discharged his weapon, the round sailing harmlessly away. Members of the section who were in the same small room were naturally startled, coming very close to injury or possibly even death. Hearing the discharge, other soldiers who were on the second floor of the Palace quickly descended the stairs to see what had happened and if anyone was hurt.

Now confronting a large body of his subordinates, most of whom were young privates, the master corporal shouted to everyone present that "you didn't see or hear anything," the obvious message being that no one was to say anything about his mistake. A corporal, who incidentally was good friends with my 2IC, made a similar statement only moments later, telling everyone to simply forget this incident. These two leaders had essentially directed their subordinates to cover up an infraction under the Code of Service Discipline ... and they complied. At this time, I was away from the

Palace on other duties related to my temporary position as the platoon's warrant officer, basically requisitioning supplies at the main UN camp in Port-au-Prince for all of my soldiers.

More than six weeks passed without this incident coming to light and then everything started to unravel, and quickly at that. One day, another section sergeant mentioned to me – I had by that time returned to my normal duties as a section commander – that he had heard a rumour concerning an accidental discharge. This particular sergeant had first heard it from one of his own soldiers. We discussed the matter as best we could, but since I knew nothing of the incident, and the other sergeant could only speculate as to what had actually occurred, we could draw few concrete conclusions. Intrigued and disturbed by this conversation, I immediately returned to my section and began to "investigate" on my own. Speaking to the master corporal, my 2IC, I asked him directly if he had heard anything about a negligent discharge. Without hesitation, he looked me straight in the eye and stated that he had not, concluding his blatant and convincing lie with a reassuring "don't worry, you can trust me." Obviously I could not. No one else in the section spoke about the incident either, preferring instead to remain silent.

Completely unaware that my 2IC had lied, I took the next logical step and approached the platoon's officer commanding, a young lieutenant, in the hopes that he might be able to uncover something about the discharge. We decided on the way forward together. I advised him that he must do something. If the lieutenant was to retain his soldiers' trust, he had to demonstrate his concern and to show that he was actively interested in uncovering the truth.[7] After a very frank discussion, it was decided that the officer would indeed address the entire platoon, which he did shortly thereafter. With the men assembled in three ranks, the lieutenant stated that the individual who was responsible for the accidental discharge could at any time come to him in private where the matter would be settled there and then (i.e., the offender would receive his punishment, which for discharges of this type in theatre was a monetary fine of $1000). The officer, to be fair, also stated that if this was just a nasty rumour with no foundation in truth, then the men were to stop talking about it at once and to continue focusing on the job at hand. As might be expected, no one came forward in the days that followed. Keeping the initiative, however, I spoke to my section, including my 2IC, and asked them in private if they knew anything about the accident. No one came forward, and for the second time, the guilty master corporal reassured me of his honesty, loyalty and faithfulness. Having exhausted all of my options, I let the matter rest. All in all, I was satisfied that the rumour was exactly that, a rumour, and nothing more could really be done.

Some time later, Haitian security forces replaced the Canadians and assumed the responsibility of providing presidential protection. Thus relieved, my section left the Palace and returned to the main UN camp where the remainder of the regiment was located. Many of my soldiers were scheduled to take some well-deserved leave at a nearby Club Med resort and so used this opportunity to prepare for their time away.[8] As things would have it, the group that was soon to depart included the offending master corporal, his friend the corporal who had earlier supported him, and the individual from the other section who had informed his own sergeant about the rumour (the one who later brought the matter to my attention.)

Soon after the group had left for Club Med, the company sergeant-major (CSM), accompanied by two military policemen (MPs), asked to speak with me in private. The trio stated that they needed to talk with my 2IC as they were investigating an assault, so serious was it that a soldier had broken his arm during the scuffle. Transportation was immediately sent to the resort, returning to the camp a little while later with the master corporal and corporal in tow. Once back in platoon lines, I asked my 2IC if he was involved in any way with this incident and, as before, he simply said "no." Although the master corporal may have fooled me, he did not fool the MPs, armed as they were with incriminating evidence. The following day he was duly charged with assault.

At this time, and for the first time, I learned the true nature of what had actually transpired. When the soldiers were preparing for leave, the master corporal visited the mess and became exceedingly drunk. He eventually returned to the platoon's lines where he found the soldier that had initially informed his own section sergeant about the rumour. Seizing upon this opportunity to exact revenge for "squealing," the master corporal took a heavy metal bar and began to hit the soldier violently, who, in an attempt to defend himself against the onslaught of blows, used his arm as a shield, thus breaking it in the process. Rather than report this incident, the now-injured soldier simply went to bed and tried to sleep as best he could. The next day, in understandable agony, he visited the MIR (medical inspection room) and sought attention for his swollen, disfigured and painful arm. Instead of recounting the truth, he simply told the doctor that he had injured his arm in the weight room. This explanation failed to convince owing to the extent and pattern of his injuries. The doctor explained to him that his infantry career was essentially over, as he would have to insert metal plates and screws into his arm so that the shattered bones might be re-aligned. Realizing that his time in the infantry, if not the military, had now reached its end, the injured soldier confessed to the doctor that he had told his section sergeant about the rumour and that he had been assaulted as a consequence. Once the MPs learned of this incident through the doctor, who was duty-bound to report it, they began to investigate the master corporal, quickly bringing legal proceedings against him. The truth had finally emerged. Duly court-martialled for his actions, the master corporal lost his rank and was sentenced to nine months confinement at the Canadian Forces Service Prison and Detention Barracks in Edmonton, at the conclusion of which he took his military release.[9]

Faced with these disturbing revelations, I immediately began to question myself. I had placed my trust in my 2IC for good reason and had been betrayed, not once, but a number of times. The master corporal had performed exceedingly well and, as a result, had earned my confidence. I had no reason to doubt him. My sense of guilt weighed heavily and I constantly wondered if I was really entitled to lead my section in light of this failure to recognize my 2IC for who he really was, a manipulative liar who intimidated his subordinates and was prepared to use force to ensure their silence. Because of my shattered self-confidence, I spoke with my CSM who explained that I was not at all to blame for what had occurred. He also related that in his long military career, his own subordinates had sometimes betrayed his trust. From his perspective, I certainly was not the first to experience such a violation of trust and I certainly would not be the last either. His advice lifted my spirits.

Having to my credit an operational tour in Cyprus and years of airborne experience, I began to realize that I was not naïve and was not wrong in placing my trust in my 2IC. As I later learned, this particular soldier was a master at deception. Quick and intelligent, he had probably always hid the truth from his superiors and intimidated his subordinates into silence. He was calculating, using his charm, wit and affability to win the trust, praise and confidence of others. So able was he that his jailors in Edmonton believed him to be a model prisoner, just as he had been a model soldier prior to Haiti! His sly approach and persuasive personality even set some of his guards at ease. I was certainly not the only one to be fooled.

In spite of the accidental discharge and assault, the remainder of my section was still compelled to fulfil their assigned responsibilities in Haiti, as was I. My challenge was now to rebuild the morale, cohesion and trust that had been weakened by this entire series of unfortunate events. Given the course of the past few weeks, I, along with members of my section, quickly realized that this was going to be an exceedingly tough and long process.

The need to maintain a coherent chain of command necessitated the replacement of the master corporal whom the MPs had arrested and charged. After considering the advice of his sergeant-major, the platoon's lieutenant selected a new 2IC for my section. Everyone, especially the new appointee, knew that he was stepping into a very difficult position, having to take over from another who had disgraced his command by violating the trust of those whom he was supposed to protect. In order to facilitate the reconstruction of trust, a definite priority, the lieutenant made a conscious decision to select a competent, well-liked and highly respected soldier. Had he chosen someone with whom others were legitimately dissatisfied, it is more than possible that they would have been reluctant to trust him given their recent experiences. I understood that my subordinate's job would be difficult, having to prove to all that he was deserving of this appointment, and so I quickly advised him to work hard, to act professionally at all times and to treat others well. If this was done, time would take care of the rest and trust would eventually be restored within the section.

Other means were also sought to repair the trust that had been lost. Soon after the assault, my section was tasked with a long-range patrol in the south of Haiti, during which we were supposed to talk to the inhabitants of three villages to determine how things were going in the area. I instantly recognized that an opportunity had just been presented that I could turn to my advantage. One evening after the duties of the day had been completed, I told my section to find the most isolated spot that they could. I detailed a member of my patrol (who also happened to be a Haitian translator) to go into the local village and purchase enough seafood for the entire section. While waiting for his return, my new 2IC and I made French fries from the potatoes that we had brought with us. At night, with only one man on guard, everyone enjoyed a fantastic meal, swam in the ocean and relaxed. As well, a Canadian national living in Haiti offered his house and grounds to members of my section, thus allowing them to spend the night in relative comfort if they wished. The time away from the other sections, the main camp and the demands of the mission proved extremely beneficial. While sitting on the beach, a number of soldiers talked to me informally about the accidental discharge and the assault, and also amongst each other, in the hopes of putting it all behind them.

During the tour, I received notification that I had been promoted to warrant officer (WO). I learned while speaking with my sergeant-major that if I accepted my crown now, I could not remain in Haiti as there was no room for an additional WO in the authorized establishment. Because of this, I would have to return to Canada, leaving my section and the mission behind. For me, abandoning everything in mid-stream for a little more formal authority was not a realistic option and so I vowed to remain a section sergeant for a while longer. I had begun the tour with these soldiers and I was determined to finish it with them as well. Once the entire regiment returned to Canada, however, I was immediately promoted. My section was in need of leadership at this time and I believed that I could not simply leave at such a critical moment. How could my men trust me later? Would they if given the chance? I knew that I had to stay.

By demonstrating concern for my soldiers' welfare, I hopefully stemmed the further erosion of confidence and started to restore trust between leaders and followers within my section. Although neither my new 2IC nor I were directly involved in these disgraceful incidents, our soldiers were naturally a little distrustful now of those in authority. We needed to demonstrate that everyone could in fact trust us, and as their section sergeant and 2IC, we would always have their best interests in mind. We had not caused the loss of trust, yet we were obligated to rebuild it. Our efforts seemed to have worked for the section continued to function effectively, just as it had prior to the accidental discharge. The tour ended two months later.

Incidentally, once the MPs became involved in the matter, some privates and corporals from my section came to seek my advice as to whether or not they should cover up the incident when questioned by the investigators. They had apparently not lost faith in me and believed that I could still be trusted. Believing that airborne soldiers do not "blade" other airborne soldiers, they reasoned that lying was perhaps the best route to follow. I indicated that it was ultimately their decision, but I advised that they consider the consequences of misleading the MPs: if they were found to have lied, they could be charged with a serious offence and the consequences would affect their wives, their children and their careers in the Canadian

Forces (CF). Much was at stake and I alluded to this fact. I also turned their own reasoning on its head. While I acknowledged that it is important for airborne *soldiers* to stick together, I argued that good airborne *leaders* never put their subordinates in a compromising position where they have to lie for them. After trying to convince them that the time for lies was over, I encouraged them to act as the professional members of the CF that they were, that is, with honour. These words must have been persuasive as they duly cooperated with the investigators, perhaps realizing that they were now presented with a means of ridding themselves of a problem soldier. Throughout the entire episode, from the accidental discharge to the regiment's eventual return to Canada some months later, the members of my section seemed to trust me still. Whatever their feelings, however, I believed that I absolutely must do something to at least reinforce trust within my small command.

In discussing the event some time later, my company commander remarked that the master corporal could have by himself ensured that this event went no further and did not escalate to the point of assault. The officer stated that had he come forward, admitted his mistake and accepted his punishment, life would have remained pretty much the same. His fellow soldiers, realizing that accidents happen, probably would have continued to respect him for acknowledging his own error and for being a "man." Instead, he tried to cover up what had happened by intimidating, and then assaulting, his subordinates. In that instant, that split second in which he told witnesses to remain silent, he completely destroyed the bonds of trust that had taken so long to establish both before and during the deployment. I agreed completely with his assessment.

Although aware of what was happening, senior leaders within the platoon largely remained in the background, preferring that I take the lead in rebuilding the trust that had been lost since it was my section that had been affected and I knew my soldiers best. The lieutenant was young and in command of a platoon for the first time. Given his relative inexperience, he relied heavily on his warrant officer and his three section sergeants, which included myself. A very good working relationship based on open and frank communication soon developed amongst us. Together, we planned how we would approach and deal with the problems created by this difficult situation. A good deal of trust existed in this small group: the lieutenant not only trusted his subordinates to provide good advice, which they did knowing that he was interested in resolving this issue, but also trusted me with implementing the agreed-upon solutions. I certainly did not want to disappoint him, which my failure most surely would have.

And so the natural questions arise: Why did everyone concerned remain silent after the accidental discharge? Why did the assaulted soldier go to bed in spite of his broken arm? *In my opinion*, the reasons were many. The Canadian Airborne Regiment had been disbanded less than two years ago in March 1995 and many were still smarting from this insult.[10] In the aftermath of this controversy, these soldiers, including myself, had worked exceedingly hard to rebuild their identity and reputation as dependable, professional warriors. When confronted with a situation that had the potential to damage their image yet again if it ever came to light, they thought it best to keep this matter quiet. These men suffered from the not-so-subtle pressure to be successful as airborne soldiers, especially since some in the military, government and public questioned their professionalism and *raison d'etre*. They were not going to let two individuals sully their name and the status of all airborne soldiers as a handful of trouble-makers had done a few years earlier in Somalia and immediately afterwards.[11] They were proud of who they were and silence seemed a logical approach at the time. Few could endure more slander and no one could afford another large (and public) mistake.

In deciding whether or not to act, other factors also entered the equation. Again, *in my opinion*, the master corporal and corporal were "bullies" and had on occasion treated their section poorly (like new recruits) by yelling at them and insisting on needless inspections. The other soldiers, mostly privates at the start of their careers, were somewhat intimidated by the pair and understandably reluctant to come forward, fearing for their safety. The later assault more than justified their apprehension. Many also believed that the master corporal was a good, competent soldier on the whole (despite his conduct at times) and they did not want to see the accidental discharge, which was one of his only errors, stain his service record. He apparently fooled them too and earned their trust on false pretences. No one had been injured either by the errant round. Above all else, however, was the belief that soldiers, especially airborne soldiers, do not "rat" on one another.[12] All of these reasons combined to keep members of the section quiet. As will be recalled, a soldier from *another* section was the only one who eventually came forward … and he paid the price for his courage, losing his infantry career in the process.

Ultimately, the trust so essential to effective military teams had been violated ... repeatedly. An honest mistake that could have easily been resolved eventually spiralled out of control and cost many soldiers their military careers and shook the confidence of others. I had not been wrong in trusting my 2IC. I had placed my confidence in him as a good leader should when working with someone who never disappointed, performed exceptionally well and was deserving of such loyalty. Only after this episode had run its course did I question my earlier actions, yet neither I nor anyone else could have foreseen what eventually occurred. So conniving was the master corporal that I had no reason to doubt that my actions were anything but correct and consistent with building a strong, cohesive and effective team. Although not responsible *per se*, I still felt obligated to rebuild trust within my section and to demonstrate to all, but especially to my subordinates, that both my new 2IC and I were trustworthy. The trust that had taken so long to establish had in an instant been destroyed and it took even longer to regain that sense of confidence so necessary to small teams and large organizations alike.

ENDNOTES

1 This chapter is based on a single interview between MWO J.B.M. Colbert and C.L. Mantle that was conducted in Kingston at CDA HQ on 19 November 2007. By way of explanation, the former described all the relevant details surrounding this incident and the latter duly acted as the pen, the "ghostwriter," essentially taking what had been provided and writing the following account in the first-person. Earlier versions of this chapter were vetted by MWO Colbert to ensure that what has been said is indeed an accurate depiction of events, attitudes, impressions and opinions. Other individuals who deployed on Op Stable with 3 R22ndR may have very different recollections of events mentioned herein, but little can be done to rectify any discrepancies in such a short piece as this.

2 "Past Canadian Commitments to United Nations and Other Peace Support Operations (as of December 2003)," http://www.forces.gc.ca/admpol, last accessed 21 November 2007.

3 For a more detailed history of Haiti from 1991 to 1997, see DND Backgrounder, UNSMIH, BG-97.007, dated 30 January 1997.

4 A more complete picture of the equipment and personnel sent to Haiti on this and other operations, including relevant background information, can be found in Canada, DND, Operational Research Division, DOR (MLA) Research Note 2004/10, D.W. Mason,

"The Canadian 'Five-W' Database," November 2004, A59-A61. Concerning 3 R22ndR specifically, two infantry companies ("A" (para) and "C") were deployed along with the service company; "B" company provided soldiers to the first two, thus ensuring that they were at full strength, with the balance remaining in Canada as the rear party.

5 Only about eight soldiers were actually available for duty at any given time, the other four being on various forms of leave.

6 Within Colbert's company, 1 platoon served at the Palace and, aside from providing perimeter security, offered both reconnaissance and counter-ambush capabilities; 2 platoon acted as the quick reaction force that could be called upon to provide additional support, security or extraction; and, 3 platoon patrolled Port-au-Prince.

7 The need to appear interested in and genuinely concerned with serious matters that can ultimately affect discipline and morale is discussed in the chapter by Major Tony Balasevicius in this volume. His chapter, like this one, also speaks of a small group of soldiers that, through intimidation and the threat of violence, held undo sway over others.

8 During the tour, soldiers were given 96 hours short leave and were allowed to choose where they would spend it; some chose the local Club Med resort because it was so close, while others travelled to Florida or the Dominican Republic.

9 As a point of note, the corporal who supported the master corporal went into a "tailspin" following this episode. Whereas before he was extremely fit, did not smoke and took excellent care of himself, once his companion had been charged he stopped visiting the gym and started using tobacco. He quickly left the CF once redeployed to Canada, taking his release before the remainder of the regiment had returned from their post-deployment leave. Colbert never learned the exact reasons behind these drastic changes (although they can be surmised) and never heard from him since. The corporal, to the best of Colbert's knowledge, was never charged.

10 For an insightful discussion of Canadian airborne forces from the Second World War to the mid-1990s, see Colonel Bernd Horn, *Bastard Sons: An Examination of Canada's Airborne Experience,* 1942-1995 (St. Catharines: Vanwell, 2001).

11 For an account of Canadian activities in Somalia and their aftermath, see Canada, *Dishonoured Legacy: The Lessons of the Somalia Affair: Report of the Commission of Inquiry into the Deployment of Canadian Forces to Somalia* (Ottawa: Minister of Public Works and Government Services Canada, 1997). Additional information can be found in David J. Bercuson, *Significant Incident: Canada's Army, the Airborne, and the Murder in Somalia* (Toronto: McClelland and Stewart, 1996).

12 For an analysis of how an over-developed sense of loyalty and an unconstrained cohesiveness can result in insular behaviours wherein members of a group act in a manner inconsistent with the conduct expected by the larger institution to which they belong, see Donna Winslow, "Misplaced Loyalties: The Role of Military Culture in the Breakdown of Discipline in Two Peace Operations," *Journal of Military and Strategic Studies* 6, no. 3 (Spring 2004), 1-18, and, Donna Winslow, "Misplaced Loyalties: The Role of Military Culture in the Breakdown of Discipline in Peace Operations," *The Canadian Review of Sociology and Anthropology* 35, no. 3 (August 1998), 345-367.

CHAPTER 3

TRUST: A CRITICAL ELEMENT OF TASK FORCE ORION

Lieutenant-Colonel Ian Hope

Editors' Opening Note: *In a previous volume of the In Harm's Way series, Lieutenant-Colonel Ian Hope described his recent operational experience while in command of Task Force (TF) Orion that deployed to Kandahar, Afghanistan in mid-January 2006.*[1] *While his chapter focused primarily on the leadership challenges of senior commanders in operations, the undeniable importance of trust appeared as a consistent theme throughout. For example, he explained:*

> *We have come to see warfare as a clash of technologies.... The high levels of journalistic prose, the mind-numbing Powerpoint and jingoism, and the elevation of far-fetched assumptions to the level of indisputable truths have conspired to let us forget that war remains, now as it was when Agamemnon met Priam, a violent clash of independent human wills.... It is in preparation for this moment that young leaders must dedicate their energy, physical and intellectual, and develop their personal value system. They must come to trust themselves, trust in the heart and head of each other, and trust in the heart and head of their superiors if they are to survive and win this violent clash.*[2]

Given Lieutenant-Colonel Hope's continual reference to trust as the cornerstone of effective operations in the new security environment, the editors decided to include as a chapter those sections of his previous In Harm's Way contribution (with minor editorial changes) that highlight the importance of trust and how his personal actions facilitated the trust-building and -maintenance process. As with all other chapters in this volume, the reader is asked to compare the following account with what has been reported in the trust literature thus far (outlined briefly in the introduction). In so doing, it is expected that one will clearly recognize how Lieutenant-Colonel Hope's actions facilitated the building and maintenance of trust during his time in Afghanistan and how this trust, in turn, lead to success in a complex operating environment.

LIEUTENANT-COLONEL HOPE WRITES ...

I had been in command of 1st Battalion, Princess Patricia's Canadian Light Infantry (1 PPCLI) for 12 months. I was first warned of our mission to Afghanistan in October 2004 and was personally told by General Rick Hillier, the Chief of the Defence Staff, in March 2005 that we would go to Kandahar under the American-led Operation Enduring Freedom. So by September, my Regimental Sergeant-Major, Chief Warrant Officer Randy Northrup, and I had created a fairly cohesive battalion team that was focusing on the upcoming mission.

At that point, however, we needed to expand the "team" (our unit) to include the other components and attachments that would make up our battle group (BG). We did this by co-locating all of our people in one bivouac, by decentralizing combat support assets and embedding them within combat sub-units, by providing training scenarios that concentrated on establishing broad trust in each arm and corps represented in the task force – mainly through extensive live-fire training – and by forcing people away from sub-unit identifiers (flags and colours) toward acceptance of a common – neutral – identifying symbol. I chose Orion from the constellations – representing the mythical Greek hunter of mountain beasts – that I knew blessed Afghan skies, so that our soldiers might look up, and seeing it, feel part of a larger entity, enduring and meaningful. I used this symbol deliberately to help create cohesion within this ad-hoc grouping. It is in efforts to create and maintain cohesion that trust becomes meaningful beyond the boundaries of the individual soldier.

... THE MISSION AND THE CONCEPT OF OPERATIONS

Our tasks were multifarious, divided into three broad categories: governance, security and reconstruction. These became our lines of operation (broad sets of connected objectives and tasks that contribute to the elimination of the key problems or obstacles to our mission). The key problem I saw was the winning and maintenance of the confidence of the people (who were at the same time subjected to considerable pressure by the Taliban to support their cause). In this sense, the struggle in Kandahar was essentially a clash of human will, between forces loyal to the government of Afghanistan and those opposing this authority, with the people of Kandahar in the middle being pulled, enticed, persuaded and coerced to choose sides.

The confidence and support of the people was essential to establishing freedom of movement and action for each of the opposing forces. We both relied upon the same source for such essential freedom and support. The people constituted a common "centre of gravity" that we and the Taliban wrestled over. When one comes to understand the conflict in these terms, one can easily see the importance of trust beyond institutional horizons. Establishing trusting relationships with local authorities and with the people was essential to swaying the centre of gravity to our side. This concept is essential to understanding the nature of counter-insurgency operations.

As an operating concept, I intended that the 1 PPCLI BG become an extension of the Provincial Reconstruction Team (PRT) so that I could coordinate and synchronize all TF tasks in a unified effort under one chain of command. This centralized control was to be exercised in a manner responsive to the desires of Afghan authorities and their people. To this end, we established a combined Afghan/Coalition Provincial Coordination Centre (PCC) at the Governor's Palace to help build the capacity of provincial civil, police and army leaders in managing security. Likewise, I wanted the PRT to operate downtown in order to work with the provincial staff to coordinate international agency reconstruction efforts and to assist in the functioning of the Provincial Development Committee (PDC). This would give the provincial leadership powerful centralized organizations that could assist them in exercising their authority while building their capacity to govern effectively. While command of the TF was centralized and responsive to the provincial authorities, the execution of the governance, security and reconstruction tasks was quite decentralized. I pushed elements of the manoeuvre sub-units out into independent sub-unit forward operating bases located in areas that would allow them to work with Afghan National Security Forces (ANSF) and district leaders to improve governance, security and socio-economic conditions in key districts of the province. We were organized in a decentralized manner and we evolved the decentralized concept to the point where there were multiple attachments within a platoon.

To enhance provincial and district governance, I wanted the TF elements to engage village, district and provincial leaders to convince them to implement Government of Afghanistan initiatives. I wanted every patrol to stop in villages and conduct local shuras (meetings) to hear out the people and to encourage the elders to engage district leaders. The patrols reported results to the officer commanding the respective sub-unit who in turn engaged district leaders to properly represent the villages to their provincial leaders. The PRT commander and myself engaged the governor and his staff and provincial council to hear representation from the districts. TF patrols were also tasked with reporting village and district reconstruction requirements to the PRT and to monitor the progress of reconstruction projects in rural areas.

... TRUSTING OURSELVES – FINDING THE ENEMY

In attempting to find Taliban fighters, TF Orion was supported by a myriad of "enablers" (Electronic Warfare (EW) and imagery platforms, and various human intelligence (HUMINT) networks). However, we could not trust in these. The brigade did not have sufficient numbers of enablers to cover every task, and our operations were normally of lower priority than any of the time-sensitive high-value target hunts being conducted by special forces. Therefore, even when we were allocated several enablers, they were often stripped away very early in an operation, leaving us relatively deaf and blind and with sinking feelings as they departed each time. When they were present, they provided useful intelligence and support, and made us feel more secure. But they were also plagued with the friction normal in combat: incompatibility of technologies; air gaps; low fidelity sensors; receivers that were not rugged enough for combat; vehicle and equipment breakdown; bad atmospherics and foul weather; restrictions on intelligence sharing that are inherent in coalition operations; and, human error. Friction degraded our sensor and communications networks to the point where they were incapable of producing more than a fleeting glimpse of a portion of the enemy's force, that is, temporary visibility upon only one piece of an immense and dynamic jigsaw puzzle.

To compensate for the lack of effective ISTAR (intelligence, surveillance, target acquisition, reconnaissance), we began to disperse across the province and live amongst the villagers in order to develop local trusts and gather intelligence through human contact. This forced the soldiers to live in the light armoured vehicles (LAVs), or beside the gun lines, often for three or more weeks at a time without shower or proper toilet, in hot, dusty, arid conditions. We lived this way beside the Afghan people, in the face of the enemy, with good psychological impact. Communication intercept (ICOM) and other EW sources often confirmed that each of our platoon, sub-unit and BG manoeuvres was having a disruptive effect on

Taliban efforts to mass in Northern Kandahar. I began to sense in March and April that, while we were seldom making direct contact with Taliban forces, we were being successful in keeping the initiative.

The maintenance of operational initiative became something tangible to me. The subordinate commanders and I could sense when we had the enemy confused or when they seemed ready to act. I began to trust intuition, especially if it complemented local HUMINT (provided by villagers or ANSF). Our concept of operations evolved to where we continuously planned and executed multiple large manoeuvre operations in order to find, fix and finish individual enemy groups in an extended effort to keep the much-valued operational initiative. Reflecting upon this now, I realize that this dynamic was seldom if ever really replicated in our peacetime training, where the ebb and flow of the initiative was only academic (probably because such things never enter a Master Events List).

In exercising the *sense* function, I relied less and less on an inventory of ISTAR sensors and more and more upon personal *sense* of where the enemy was, what he intended to do and how he could be deceived. TF leaders personally reconnoitred districts and analyzed local HUMINT sources to identify trends and to guess if the enemy was in a general area (almost always this was relayed to us as a 20-40 man Taliban group hiding in the vicinity of a particular village). We then attempted to manoeuvre into that district quietly under cover of darkness using deception and, as much as possible, isolate the village by using thin blocking and cut-off forces. We would conduct manoeuvre (cordon and searches) and fires (show of force artillery or 25mm fire) to produce enemy ICOM chatter, and from this ICOM chatter (or HUMINT from local nationals), we would attempt to vector in upon the enemy's locations. However, in the end, finding him was almost always a result of advance-to-contact in the close country where he hid and was confirmed by the exchange of fire at close-quarters.

Intuition and trust were central in our efforts to find Taliban forces. Over time we began to decipher which Afghan information sources were reliable (and which were not) and to trust our instincts, regardless of whether or not they were confirmed by our own ISTAR capabilities. I began to trust the subordinate commanders to act – or not – upon this local HUMINT. Given the lack of "actionable intelligence" about the enemy, there were few specific tasks given to TF Orion by the brigade, so it was up to us to establish tempo and task. Operational tempo and effectiveness were almost entirely dependent upon the personalities of the commanders. We could have very easily hunkered down into localized routines in relatively safe areas, following a pattern of "framework patrolling" and by this be seen as "busy," when in fact we would achieve little (except to surrender the initiative and set ourselves up for attacks by improvised explosive devices (IEDs)). Instead, I forced myself and encouraged – and trusted – the company commanders to be perpetually proactive in hunting the enemy, forcing everyone to live "outside the wire," to continuously manoeuvre and to engage locals. This was difficult for everyone, especially after casualties were incurred, when there was always a natural tendency to adopt caution and reduce tempo. It was at these times especially that I trusted the subordinate commanders to demonstrate an aggressive spirit and maintain operational initiative through offensive action. Not all people are equal in this regard, and the army must learn that robustness and determination are not nice-to-have qualities in manoeuvre commanders; these are core qualities, and we must work hard to select the few who have the tenacity to sustain a high level of resolve in the face of danger.

... TRUSTING EACH OTHER – FIXING AND FINISHING THE ENEMY

Once found, we would attempt to fix the enemy with fires and destroy him in close-quarter combat. Neither of these things was easy. It required us to stay within 100-150 metres of the enemy and to coordinate fires before physically moving to clear his positions. Doing so was contrary to human nature. Like operational tempo, the degree of success we had in close-quarter combat became personality dependent. The majority of soldiers when fired upon for the first time would seek to disengage back toward the "last safe place" they occupied. After several encounters, they repressed this urge but would be very reluctant to advance in contact (especially when separated from their LAVs). Forward movement or staying in place on the close-quarter battlefield (especially after night-fall) depended upon the continued presence of battalion and company commanders, supported by the "natural fighters" in our ranks. It became evident to me in May that the number of true fighters we had was a small minority. By "fighter," I mean those men and women predisposed to keep fighting regardless of violence and danger; those who repressed fear not just because they wanted to remain with their primary group, but because of an overwhelming desire to beat the enemy; those who truly wanted to hunt the enemy and make him the victim. I would estimate that there were only six or seven such individuals in every 40-man platoon. Yet, their stalwartness almost always became the psychological pivot point for the action of a section, platoon or company engaged in intensive battle. I came to rely upon the courage of commanders and this small number of fighters in each platoon and company to override the inherent fear of close-quarter battle and to ensure that we kept the enemy fixed before closing to destroy him. Trust – not authority – was the essential element.

It became silently understood that trust and cohesion required an equitable sharing of risk. I travelled everywhere by vehicle, slept on the ground and avoided using helicopters because of the impression this left upon the troops. Instead, I would personally lead all road moves until the point where I felt we would begin a final advance-to-contact. It is my philosophy that the commanding officer must command the lead vehicle, must navigate and make the tactical decisions during all phases of movement. This served several purposes: it alleviated a subordinate commander of an additional task of route navigation prior to a battle (allowing him to focus on the close fight after he was delivered to it); it took advantage of the most experienced crew in the BG, which traveled over 9,700 kilometres during the tour and knew the routes and terrain better than any other (this helped avoid unnecessary navigation errors and personal embarrassments to junior leaders who needed their confidence intact before going into battle); and, it demonstrated a sharing of risk. The approach came with cost: the LAV in which I travelled was hit four times by IEDs and rocket propelled grenades, resulting in the loss of two of my crew, Master Corporal Loewen and Master Corporal White, grievously injured on separate occasions by shrapnel. This demonstration of shared risk went far beyond any authority granted by rank and appointment to promote the trust essential to the fighting spirit of TF Orion.

The company commanders were also always seen well forward in the fighting. They were entrusted to personally lead the close-quarter fight and to assist me in my role of coordinator of the close battle. I trusted and relied upon these commanders to apply the determination, fierceness and courage that this hard task required. In turn, the junior leaders could rely upon their superiors to provide the best coordination of support.

In the coordinating function, I found that the evacuation of our dead and wounded, and the recovery of our damaged LAVs, to be of the highest importance and I was prepared to stop all forward advance in order to get these things correct. This was different than what we had taught and practiced in training, where we preached that we would not let casualties slow down forward momentum, having second-in-commands and company sergeants-major sort out casualties while commanders continued to attack. The reality of wounding and death changed that notion. Commanders became personally involved in evacuation. It became our number one concern in battle, with positive effect. Soldiers came to understand that if wounded or killed, their recovery from those hellish close-quarters was the first priority of the chain of command, beyond any "prize" of Taliban defeat. We did not have to articulate this, it came naturally in mutual understanding and, I believe, in trust.

The idea that the commanders of TF Orion wanted the soldiers of the TF to fight, and by destruction to win, was fundamental to troop morale. Over time, the soldiers came to know that their commanders were constantly looking for enemy for them to kill or capture and had trust in their ability to do just that. During the period of April-May, "A" Company had the misfortune of operating in the mountains of Showali-kot, where the enemy avoided contact, but made maximum use of IEDs. Morale in the company was affected. To reverse this, and to demonstrate trust in the company, I ordered them into the fight in Zharie/Panjwayi and Helmand in July, where they demonstrated their proficiency in arms and where they exacted retribution upon the Taliban. Their morale soared.

... MISSION COMMAND

Task Force Orion operated under the principles of mission command, the philosophy that encourages decentralized decision-making, freedom and speed of action and initiative in support of command intent.[3] Successful mission command is hinged upon mutual trust, where commanders have confidence in their subordinates to exploit opportunities and complete tasks and where, in turn, subordinates trust that their superiors will support their decisions. Most often the companies operated independently, with the command LAV moving between them fulfilling coordination functions. Frequently, companies would fight disparate engagements. At the point of highest tempo in July, the TF fought simultaneously in dispersed operations in Panjwayi, Sangin and Hyderabad, then Panjwayi, Nawa and Garmser (each company fighting 50-100 kilometres from the others). Superb sub-unit leadership, supported by outstanding operational coordination by a brilliant BG staff, handled all these separate fights. At no point in this period of prolonged dispersed combat did I have knowledge of the tactical situation in each sub-unit fight, nor did I need it, since I trusted each sub-unit to continue to execute our concept of offensive manoeuvre operations designed to reduce enemy capacity, maintain the initiative and continuously create conditions that I could exploit positively in information operations.

This is not to say that I did not expect mistakes to be made. All the commanders of TF Orion made mistakes in operations, mostly because the experience of sustained combat operations was new to us and we had no paradigm or experiential guides. Tactical errors were made that cost life and limb. We who were responsible for these decisions and who gave these orders had to deal with that knowledge immediately – in the face of continued operational pressures – without compromising the faith and trust necessary for mission command to work. Task Force Orion instituted learning processes to capture lessons learned after each major engagement and to disseminate these throughout the area of operations. For my part, I emphasized use of the after action review at all levels, sometimes running them myself, in the field, after the fighting was over.

In all of this we did not dwell upon mistakes. Sometimes we did not even speak of them. Nothing is more damaging to trust and self-confidence than to have a man who acted decisively in combat, but was the victim of bad luck, be interrogated by a commander who was not present and had not felt the pressures and complexities of that moment. While I desired that commanders determine quickly what had gone right and what had gone wrong, I did not want this to create an environment of blame. Dealing with loss was a powerful enough motivator to quick learning. Finding quick forgiveness of one another went a long way to avoiding the trap of blame games.

... DISCIPLINE, TRUST AND COHESION

Mission command was sustained because trust had been established well before the fighting started with a unit culture that promoted self-criticism and self-adjustment. I deliberately steered away from the Canadian Forces (CF) tendency to order investigations. When we were asked our opinion of participating in standard investigations into all combat actions, I baulked at the idea for two reasons. First, no investigative group from Kandahar Airfield or Canada was qualified to understand or replicate the circumstances of the combat we faced. Second, an automatic default to outside investigations was an automatic statement of distrust in commanders who for the first time in half a century were dealing with the unknowns of combat. My response to initial attempts by the CF to investigate every firefight was simple: if we look to blame someone for any of this, let it be first understood that when a man is wounded or killed on a battlefield, there is a lot of blame to go around. War is messy, confusing and unpredictable. Predictability and certainty are never achieved. Determining cause and effect is extremely difficult, if not entirely impossible. To attempt it as a bureaucratic standard would seriously erode trust.

I am not suggesting in these statements that we tolerated breaches in the Code of Service Discipline or the Law of Armed Conflict. If I sensed that these were at issue, there was only one recourse: swift and decisive action. Discipline is an absolute essential in combat units and clear boundaries must be drawn, expectations must be announced and disseminated, and transgressions must be firmly dealt with. This extended down to the seemingly trivialities of dress. I allowed soldiers to wear whatever load-carrying apparel, boots and gloves they wished. I demanded that they wear proper PPE (personal protective equipment) and IFF (identification friend or foe), and took swift action when they did not. I would also not tolerate the proliferation of platoon or company badges or flags, as they jeopardize proper PPE and IFF, and because they reinforce sub-unit cohesion at the expense of unit cohesion.

It was beyond unit boundaries that things became more problematic. The investigative culture of the garrison CF was antipathetic to our work. We received orders to pass to the military police responsibility for ALL investigations for accidental and negligent discharges. This delayed the expediency of summary trail and left soldiers anguishing unnecessarily. It also served to diminish the authority of the chain of command. At other times, CF-sponsored investigations were conducted that I was not made aware of; these were synonymous with votes of non-confidence. These things breached the principles of mission command and of a trusting command climate. To help reduce the negative impacts and to reinforce the role and trust of commanders, I instituted a policy wherein any discharge of weapons outside of firefights – from either negligent discharges or escalations of force – required the company commanders to inform me in writing within 24 hours of the who, what, where and when of the incident and to state whether or not they believed the incident required investigation for possible breach of the rules of engagement. I gave consideration to these and would act appropriately. As this policy required automatic reaction and direct involvement of company commanders, it served to reinforce their role and to make us accountable for knowledge of subordinate actions. In this manner, to a small degree, it helped reduce the garrison investigative mentality.

... TRUST, FEAR AND LOSS

Risk of injury or death was shared, and I, and the sub-unit commanders, came to know intimately the fears of moving across that imaginary line that separates what constituted safe ground and what we knew to be a dangerous place. We came to understand why soldiers kissed their personal talismans before a fight, the fatalistic acceptance of death, how to deal with that visiting shadow that popped the dreaded question into my head, "will I ever see my children again," and how to dismiss fears and carry on with our tasks. These psychological oscillations are never present in peacetime training; they were the constant companions of those who lived outside the wire and mark them permanently from those whose relative safety precluded having to deal with that dark shadow. What most visiting officers failed to grasp when talking with our soldiers was that the soldiers had begun to judge their leaders by how much personal risk they were prepared to assume. Helicopter "drive by" visits by senior leadership who left to go and sleep in comfortable places never left any impression upon the soldiers. The commanders left in the field began to understand that to have these soldiers respond right in a fight meant being present – up front – and demonstrating that rank made no one immune from the awful consequences of combat. Shared risk is essential in establishing hierarchical trusts.

To sustain a unit in prolonged combat, there must be a great deal of hierarchical trust. That trust demands understanding and forgiveness, especially when it is clear that fear is visiting. I did not question any request by a subordinate commander who asked for a long water break or who needed time to sort out communications issues before or during battles. Sometimes I, or one of my subordinate commanders, needed a moment to thrash the dark angel that had chosen that moment to visit, before picking up and carrying on. I knew that commanders were sometimes dealing with battles within themselves, but I never asked the embarrassing questions. To let a man know face-to-face that you see fear in him is to forever compromise him to you. To let their subordinates know that you have seen this fear in their commander would spread the fear far. Better to grant him time to deal with it and make it clear that you expect action consistent with the maintenance of the initiative over the enemy. With time most of us became familiar with the violence and uncertainty to a point where right-brain control (rational thinking) combined with fierce determination could win over fear. But it takes time and exposure to come to this point, and trust and understanding and patience are applicable in learning to fight.

If a soldier or leader was known to many to have faltered in combat from fear, there was only one recourse. The commanding officer, if he has any trust left at all, must place these individuals beside him and prove to the collective that they are being granted a chance to redeem. They will not let him down.

... CONCLUSION

After our first trials in combat, once we had settled down from first excitements, I trusted in the subordinate commanders because they had natural tenacity and robustness. I trusted that they would not succumb to their fears, but would combine their talent with the natural fighters under their command to enable the platoon or sub-unit to sustain and advance in the midst of chaos, violence, uncertainty and fear to fix and finish the enemy. I trusted that they would learn from mistakes and successes alike, that they would handle death and loss quickly, with dignity, and then move on. I trusted that they would enforce good discipline and order, and yet exercise forgiveness and compassion, without jeopardizing trust and lapsing into blames. All of these trusts, when combined, produced the climate of mission command necessary to winning a counter-insurgency. None of it required sophisticated technologies. The emphasis the Canadian army placed upon "big head, small-body," "intelligence-driven," "command-centric" operations disappeared in the lush green and highly compartmentalized terrain of Zharie/Panjwayi where good leadership based upon intellect, imagination, experience, determination and trust produced the effective destruction of the enemy. Above all, it was the constituent elements of trust that won. As we move forward in army and CF developments, may we never lose sight of this again.

Editors' Closing Note: *From the outset, Lieutenant-Colonel Ian Hope has clearly identified trust as a critical element of mission success. During his time in Afghanistan, and while preparing for it, he recognized the need to build trust early and accomplished this through identifying a common purpose and identity within Task Force Orion. His actions – sharing risk and experiences with subordinates, constant personal contact with all, supporting the decisions of subordinate commanders, maintaining open communications, demonstrating competence, eliminating trust-eroding behaviour – served to facilitate the creation and maintenance of trust over time. He would most likely agree that trust writ large is indeed an essential and powerful force multiplier.*

ENDNOTES

1 Lieutenant-Colonel Ian Hope, "Reflections on Afghanistan: Commanding Task Force Orion," in Colonel Bernd Horn, ed., *In Harm's Way – "The Buck Stops Here": Senior Commanders on Operations* (Kingston: CDA Press, 2007), 211-226.

2 Ibid., 211.

3 For an additional discussion of mission command, see the chapter by Colonel Bernd Horn in this present volume.

CHAPTER 4

BUILDING TRUST IN A MULTI-NATIONAL TEAM DURING COMBAT OR NEAR-COMBAT OPERATIONS

Major Brent Beardsley

From July 1993 until May 1994, I had the privilege of serving as the personal staff officer to Canadian Major-General Roméo Dallaire, Force Commander of the United Nations Assistance Mission for Rwanda (UNAMIR), before and during the genocide that swept that nation in 1994 and resulted in the murder of 800,000 largely innocent men, women and children in approximately one hundred days.[1] Together, we *were* the Canadian contingent until late April 1994 when Canada began to deploy more personnel to Rwanda. Our contingent eventually grew to several hundred after the genocide.

During the time that I served under General Dallaire, he not only earned my respect, but also my trust and the trust of my comrades in UNAMIR. The aim of this chapter, therefore, is to explain how he earned and maintained this most essential of attributes, with a view to providing insights into the complicated nature of building trust within diverse military teams under extremely adverse conditions. During the dangerous and at times hopeless days of April and May 1994, trust was the only "glue" that held our team together as we attempted to respond to a genocide and civil war with inadequate international support.

In a previous volume of the *In Harm's Way* series, in an chapter entitled "A Profile of Effective Officership in Multi-national Command," I provided a discussion of how the UNAMIR mission was blessed with a group of outstanding senior international officers, who with their individual and collective values, skills, knowledge and attitudes, in addition to their sense of duty and their phenomenal moral and physical courage, directly contributed to saving the lives of over 40,000 Rwandans. With virtually no political will behind the mission and with inadequate resources to do the job, they achieved this success with a small force of 444 soldiers representing 21 nations. Truly, they provided the only positive effort by the world in response to the genocide in Rwanda.[2]

How was it that such a diverse force, under such adverse and horrendous conditions, in the middle of the Rwandan genocide, with virtually no moral, political or physical support from participating nations or the international community of the United Nations (UN), could achieve so much with so little? How was it that such a diverse force continued day after day to risk their lives to save Rwandan strangers, while the world and their home countries looked on as apathetic and hypocritical spectators who completely failed the people of Rwanda in their hour of dire need? I believe in many ways that the answers to these questions are directly due to the reciprocal trust that was earned and maintained between leaders and followers at all levels in UNAMIR during the genocide.

TRUST

My personal definition of trust is "the firm belief in the reliability of a person, persons or organization." A necessary condition for any military organization to be effective and to achieve mission success under adverse or combat/near-combat conditions is reciprocal trust between leaders and followers. Soldiers will not risk their lives for the leader of a mission in whom they have no or limited confidence. Trust is the glue that holds a military team together through the situational, moral and physical effects of fear, fatigue, injury, death, adversity, suffering and sacrifice. With trust, a military team can achieve much more than any leader could reasonably expect. Followers who trust their leaders will give their best to that leader, to their team and to their mission. Without trust, a military team will never achieve its maximum potential. While loyalty and obedience can be demanded by law and custom in any military organization, trust can only be earned and it must be continually re-earned. Followers may obey and be loyal to a leader because they have to, but they will only trust a leader because they want to. Therefore, how can trust be earned?

From my experience in UNAMIR in 1993-1994, under the leadership of General Dallaire, I learned that a leader primarily earns and maintains trust through three personal qualities, each of which must be recognized and admired by followers in all that a leader says and does (i.e., the walk and the talk). These qualities include competence, character and intellect.

COMPETENCE

Competence speaks to the expertise and the change capacities of the leader from the tactical to the strategic level as appropriate to their rank and appointment. For leaders to be trusted by their followers, they must consistently demonstrate that they have mastered the skills, knowledge and abilities required to perform their duties effectively. Followers will not trust a leader who is clearly incompetent. They may forgive a gap in knowledge that is clearly beyond that which is normally expected (e.g., handling of a new piece of equipment), but they will expect their leader to rapidly assimilate the knowledge or skills required to competently forge ahead. The responsibility to achieve, maintain and enhance competence lies with the leader and can be accomplished through training, education, experience and/or self-development. When leaders realize or determine that a gap exists in their skills, knowledge and/or abilities, they must determine how they are going to bridge and close that gap. They must do it. Incompetence cannot be excused and should not be tolerated. When the lives of men and women, Canadians and others, including innocents, are on the line, incompetent leaders will get people killed and will directly contribute to failure.

During the mission in Rwanda, General Dallaire, and indeed all of our senior commanders, displayed incredibly high standards of competence. This competence was demonstrated through basic soldier skills (e.g., driving, navigation, first aid, weapons handling and radio voice procedure) and higher or more complex skills and competencies (e.g., negotiation, mediation and operational level planning).

General Dallaire was fully prepared and qualified to command a multinational force at the tactical and operational level under adverse, combat/near-combat conditions. His competence was repeatedly demonstrated to all members of the mission through the superb quality of plans, clarity of orders, effective supervision and conduct of operations, and in the follow-up learning that was so essential to capture lessons learned. Every mission task may not have been a success, but there was widespread confidence that all tasks at all levels were competently prepared for and conducted to optimize the chances of success. Friction, the fog of war, inadequate logistics support, restricting rules of engagement and the absence of a sustainable combat capability, all factors above and beyond the control of the commander, can frustrate the best laid plans, and they certainly did in Rwanda.

General Dallaire's high standard of competence was clear from the outset. On one occasion, I found him giving a weapons lecture on a M72 rocket launcher to a young Ghanaian soldier who had been given this weapon without any training. He frequently drove his own vehicle from the point of the convoy and took full responsibility for its navigation. On the radio nets, his voice procedure was calm, clear and concise, inspiring confidence and focusing all personnel on the tasks at hand. In the headquarters, I frequently observed him supervising the planning and conduct of tactical operations, while also preparing an operational level reinforcement plan and simultaneously providing critical input into strategic plans and policy decisions that were being made at UN Headquarters in New York. He could work seamlessly between the tactical, operational and strategic levels of war. In the field, his personal field skills never caused embarrassment and could inspire and impress. In the headquarters, his confidence, training and experience in command and control permitted the staff on at least three occasions to be totally rebuilt from scratch (as a result of contingents withdrawing from the mission), while simultaneously continuing to plan and conduct operations.

Followers need to know and see that their leaders are competent and possess the necessary skills, knowledge and abilities to an expert level. This is the foundation for trusting their lives to that leader. Competence is best demonstrated during periods of violent change, in addition to the effectiveness and expertise the commander can bring to bear in anticipating and successfully responding to changes in task, the mission or the operational situation. Through consistent and continual demonstrations of his high standard of professional military competence, commensurate to his duties and responsibilities as a soldier and as a force commander, General Dallaire earned the trust of his subordinates.

CHARACTER

Character speaks to the professional military ethos and social capacities of the leader. UNAMIR was made up of some 444 officers and men from 21 different nations. Beyond the fact that we were all men and soldiers (and even that was not a common factor as we had significant numbers of naval and air force officers and a wide diversity of cultures), there was NO common operating language, culture, training, education or experience. With such a diverse organization, how does a leader build trust? The simple answer, and the second essential quality, is character. Character is the personal values, beliefs and attitudes that drive behaviour in speech and action. General Dallaire and our other senior leaders were men of character and this character inspired trust and confidence in their followers.

Character may be based on many components, but the most important are integrity, a strong sense of duty, loyalty and courage, both physical and moral. Integrity speaks to the honesty of the leader. Duty speaks to a focus on mission success. Loyalty speaks to being true and faithful to superiors, peers and subordinates alike. Courage speaks to the ability to overcome fear and to do what is right and required. Followers, regardless of their nationality or cultural background, need to hear and see their leader's character. Trust will be achieved if leaders consistently demonstrate their character in speech and deed.

In Rwanda, General Dallaire was always totally honest with his subordinates in that he never hid or sugar-coated our dilemma. We had little to no hope of physical support for our mission in the form of sufficient water, food, fuel, ammunition, defensive stores, general and technical stores, spare parts, casualty treatment or evacuation, etc. He never hid the fact that there was no cavalry over the hill; no one was coming to reinforce or evacuate us from Rwanda. We had been abandoned by the sponsoring organization, by our countries and by the international community. Our future was uncertain at best. He did ensure, however, that what little we had was ruthlessly shared with all, and that while our physical needs were seldom met, he ensured that we understood the moral and intellectual reasons for this mission and that risking our lives was *just* and *important*. Doing everything within our power and capability to save the lives of as many innocent Rwandans as possible and to offer some hope to the millions who had been abandoned by the world was the right thing to do and worth the risk to our lives. In sum, to demonstrate integrity, a leader must: maintain constant communication with followers; consistently set the example of sacrifice; support followers; be seen to share every risk and threat that followers face; and, continually inspire followers through personal example. General Dallaire did all this … and more.

A strong sense of duty is essential in building trust. The commander has to assign the mission and tasks, articulate a vision of an end-state and focus every follower and every resource on achieving mission success. The mission has to be just, important and achievable. In the case of Rwanda, General Dallaire had to inspire confidence that the mission was achievable to such a degree that followers were prepared to risk their lives because they were convinced it was right, important and achievable. He took every opportunity, almost daily, to visit the outposts of his command and speak directly to leaders and followers to enthuse them with his sense of duty to the mission. His persuasive communication skills and willingness to share risk in order to speak to and be with his troops inspired them to continue prosecuting the mission even when hope of support was slim and the future was bleak.

Loyalty to superiors (i.e., the UN in New York), peers (i.e., subordinate commanders) and subordinates (i.e., UNAMIR personnel), was consistently demonstrated in word and deed. Loyalty to the UN was severely tested as our parent organization largely abandoned us in Rwanda. It would have been easy to be disloyal and blame the UN for any and all failures. Amazingly, General Dallaire never once voiced to his followers any disloyalty. Whether he liked or disliked, agreed or disagreed with those orders, or believed the direction was right or wrong, he consistently acknowledged that the UN was our lawful authority and that our common duty demanded that we obey lawful commands from our lawful authority. This was our only ethical, moral and legal course of action. In private conversation with New York, he challenged many if not all of the orders that we were given, but when the decision was made, he loyally obeyed the direction received and did the best that he could to see that it was fully implemented. While some have criticized him for obeying orders that they believe were immoral, General Dallaire correctly recognized that a soldier must obey all lawful commands from a lawful authority. This is the only course to ensure that we remain soldiers and not a band of brigands.

Loyalty to peers was demonstrated through continuous dialogue with his top commanders and by being open to constructive criticism and suggestions. General Dallaire developed a very close and frank, but always respectful, relationship with all of his subordinate commanders. This gave them the ability to influence operations and to provide their input, but in return, they fully complied with all final decisions.

Loyalty to subordinates was severely tested in Rwanda. General Dallaire knew that he could not provide even the basic level of physical support to his troops and that he was demanding that they risk their lives for a cause that, while just and important, was not even minimally supported by the international community. In demonstrating his loyalty to his followers, General Dallaire consistently shared their conditions and shortages. Leadership by example is the ultimate way to demonstrate loyalty to subordinates. Sharing the dangers, risks and threats, living under the same adverse conditions, denying self for the benefit of others, and consistently communicating both the importance and justness of the mission were all ways in which General Dallaire gained and maintained the loyalty of his subordinates. In addition, fighting with superior headquarters for support, representing his subordinates' views and condition to those in authority, and always giving the credit for victories, no matter how small, first and foremost to subordinates, while publicly and solely taking responsibility for any and all defeats, established loyalty and in turn helped to build and sustain trust.

Courage is the final aspect of character. The CF has traditionally praised moral courage, because devoid of combat operations, physical courage was seldom demanded. In combat or near-combat operations, the leader must consistently overcome his personal fear to effectively perform any duty demanded by the situation and to lead by example regardless of the risk to life or limb. He must expose himself to the same dangers and risks that he demands of his followers. On a daily basis, General Dallaire exposed himself to the same, if not higher, levels of danger than his soldiers experienced. He consistently moved throughout the war zone in the lead vehicle and ventured to any point of danger where his troops were engaged. By sharing their situation and by remaining calm, confident and effective, he earned the respect and trust of his followers. In turn, they reciprocated his example through numerous acts of selfless physical courage, which on several occasions resulted in their death or serious injury.

In addition to this physical courage is of course moral courage, the courage to support what is morally and ethically right and to oppose what is morally or ethically wrong, regardless of the consequences. General Dallaire demonstrated this moral courage when in the early days of the genocide he refused to accept the withdrawal of the mission, as it would have meant abandoning and exposing tens of thousands of Rwandans, under our protection, to certain death. Right up the chain of command from our political head of mission (Special Representative of the UN Secretary-General), through members of the UN Security Council, to senior leaders in the UN Secretariat, and ultimately to the UN Secretary-General, he refused to consider full withdrawal of the mission while there was hope to end the genocide and protect and save as many Rwandans as possible. This action was not played out in the media, with his subordinates, or in any other open, self-serving manner. It was done behind closed doors, on a phone or in reports. He used his moral courage to win the day and retain a skeleton UNAMIR on the ground in Rwanda during the genocide, which ultimately saved the lives of at least 40,000 Rwandans and provided the only beacon of decent behaviour on behalf of the international community, which at best displayed despicable cowardice. Without question, General Dallaire consistently demonstrated physical and moral courage in the face of danger and evil. In all, the character of a leader is an essential quality and component of building and maintaining trust within the team and this character must be based on transparent, sincere and high standards of integrity, a strong sense of duty, loyalty and courage.

INTELLECT

Intellect speaks to the cognitive capacities of the leader. No follower wants to work for a "dummy." Leaders must be bright. While training provides standard responses to known situations, education provides reasoned responses to unknown or unpredictable situations. Experience reinforces both of the above and self-development fills the gaps in education, training and experience. Education develops not only reasoning skills, but intellectual or cognitive capacity. Leaders must not only exercise proficient analysis, knowledge and judgment, but they must also be blessed with the cognitive skills, knowledge and abilities to think creatively in new, unforeseen or unknown situations. They must be comfortable with the abstract, and if necessary, be prepared to create new knowledge. They must be masters at reasoning their way through those situations for which they have not been formally prepared and for which there is no policy, doctrine or simple straightforward solutions.

Followers need to be able to respect their leaders as being creative, abstract and even systems thinkers who can rapidly assimilate, process and apply new information to successfully conduct new tasks in unpredictable situations and environments. Intellect goes far beyond the skills, knowledge and abilities of simple training. It is a life-long process of acquiring and applying new knowledge and skills through formal and informal academic and military education, experience and life-long self-development.

General Dallaire was particularly gifted with a fine intellect. He was well educated both militarily (Staff School, United States Marine Corps Command and Staff College and United Kingdom Higher Command and Staff College) and academically (a BSc and some graduate work). He was also well read in a wide variety of disciplines (primarily history, philosophy and project management), possessed a strong work ethic, and had a thirst for new knowledge. Most important, he could recognize these traits in others and he could place bright intellects in key positions to support and implement key decisions and plans.

There was no theory, policy or doctrine for combating genocide and ending a civil war with an operationally ineffective Chapter VI peacekeeping force. At the time, there was no precedent, case studies, "book" or proven methodology (tactics, techniques or procedures) for dealing with genocide. General Dallaire, his subordinate commanders and a group of intellectually gifted soldiers had to reason their way through the friction, fog and chaos of the situation in Rwanda. Their moral compasses and intellects were their major weapons, which they had to bring to bear in a legal, ethical, moral and effective manner to prevent the genocide. No effective policy, effective direction, clarity of purpose or vision of a common end-state were provided to the commander. General Dallaire not only had to command, he had to operate and thrive in a situation of ever changing directionless chaos. In this new security environment of the post-cold war world, only those gifted with a brilliant intellect, like Dallaire and others, will be able to succeed.

Followers, through listening to their leaders, watching their performance and judging their actions, will begin to gain confidence in the intellect of their leaders, especially when they are entering new, dangerous, ever-changing and unpredictable situations. Confidence in this critical ability of the leader is essential to gaining and maintaining the trust of followers.

CONCLUSION

Using examples drawn from the UNAMIR mission of 1993-1994 and the personal example of *Major-General Roméo Dallaire*, I have provided an examination of trust. This discussion has stressed the importance of this essential attribute within a military team that is operating under adverse, combat or near-combat conditions. It has also examined the critical components of competence, character and intellect in building and maintaining trust.

In conclusion, while trust is a two-way requirement between followers and leaders, the latter must first and foremost give it unconditionally to the former. Followers must believe their leaders trust them. Trust may be difficult to build, but it is easy to lose. Once lost, it will take much time and effort to rebuild. Leaders at all levels must therefore observe, study and understand how trust operates. Leaders can build trust by unconditionally placing trust in their followers, taking every opportunity to develop their personal competence, character and intellect, as well as that of their subordinate leaders and followers, and by always prioritizing the building of trust within their team. While mission success cannot be absolutely guaranteed based solely on trust, it certainly will never be achieved if reciprocal trust between leaders and followers does not exist. Attention to this critical glue that holds a team together must be a primary concern of leaders at all levels.

ENDNOTES

1 Gerard Prunier, *The Rwanda Crisis: History of a Genocide* (New York: Columbia University Press, 1995), 261-265. Prunier was the first scholar to investigate the number of victims of the Rwandan genocide and concluded that the number was between 800,000 and 850,000. Human Rights Watch commissioned a later and more extensive investigation that erred on the side of the conservative and concluded that no less than 500,000 were killed in the genocide. There is still considerable debate amongst scholars and activists on the exact number that were killed and the conclusion of many is that we will never know for sure. See Human Rights Watch, *Leave None to Tell The Story: Genocide in Rwanda* (New York: Human Rights Watch, 1999), 1. The Report of the International Panel of Eminent Personalities to Investigate the 1994 Genocide in Rwanda, *Rwanda: The Preventable Genocide* (Addis: Organization of African United Press, 2000), 80, supports this conclusion and sets the number at no less than 500,000 and more likely up to 800,000. The current Government of Rwanda often quotes a figure of over 1 million. In spite of this confusion, or perhaps because of it, 800,000 has become the most generally cited number of victims of the Rwandan genocide.

2 See Colonel Bernd Horn, ed., *In Harm's Way – Serving the Greater Good: Perspectives of Operational Duty* (Kingston: CDA Press, 2006), 37-54.

CHAPTER 5

OPERATION HARMONY AND VARIOUS DIMENSIONS OF TRUST

Master Corporal (Ret'd) Shawn Raines

> Learning leadership from a good leader is important, but one should not be too quick to ignore the methods of a poor leader. There are many things to learn from a poor leader. It will remind you how not to respond to the needs of both superiors and subordinates.
>
> ~ The Author.

The Concise Oxford Dictionary defines trust as "faith or confidence in the loyalty, veracity, reliability, strength, etc., of a person or thing; the obligation or responsibility placed on a person who is trusted or relied on; reliance on the truth of a statement without examination; and a thing or person committed to one's."[1] These are heady words that in today's society have the same weight and carry the same resolve as "sorry" or "in a minute." In reality, however, trust is constantly and consistently violated as a matter of practice. More often than not, it has been my personal experience that little thought is ever given to one's actions or inactions in circumstances where trust is paramount. One need only look at the unconditional love and trust in children's eyes as they look at their parents. There is no deeper trust that is granted and no greater trust that can be broken or violated.

In the same vein, from the day that soldiers take the oath, we have committed ourselves to a structure that is built on trust. We trust that our immediate supervisors will have our best interests at heart. We trust that the system will provide for us when called upon. We trust that our leaders will keep us safe when they can, recognize us, and look after our families when we cannot. Sadly, for the most part, these trusts are not given the necessary diligence and attention that they require.

In hindsight, there is no doubt that Rotation 0 of Operation (Op) Harmony, United Nations Protection Force (UNPROFOR), which went into the former Republic of Yugoslavia, was a significant undertaking. Having been posted to

Germany, moving into the war zone was merely a train ride away for us. Looking over the photographs and videotapes of the departure, everyone appeared keen and eager to move into theatre and to serve as peacekeepers. We were augmentation staff from 4th Service Battalion (4 Svc Bn) to 1st Battalion, Royal 22nd Regiment (1 R22ndR).

I was young (having had my 25th birthday in the middle of the mission) and ready to do my bit. I was on my first UN mission and had always listened to the "old guys" with wide-eyed enthusiasm when they talked about Cyprus and the Golan Heights. Six months of extra pay, being away from the daily grind, and a chance to wear the blue beret! I am not sure if it was trust or ignorance that led us into a false sense of security, but our mettle would soon be tested time and again. Within 30 minutes of being on the ground, the train that I had just arrived on was shelled with artillery. It was immediately clear that this was not Cyprus or the Golan.

THE FIRST VIOLATION OF TRUST

> To lead uninstructed people to war is to throw them away.
>
> ~ Confucius, 480 B.C.

We expected that our leaders, our commanders and everyone up the chain of command, including the United Nations (UN), would not throw us into a situation unprepared, under-armed and without any realization of what to anticipate. We had faith in our government and leaders that we had all the pertinent foreknowledge and information about the situation in Yugoslavia. We trusted that we had good intelligence and that there would be no big surprises. That trust was shattered within minutes of our arrival.

It is hard to imagine, but on this mission, pre-deployment training was hit and miss, getting kitted-out was a struggle and rules of engagement (ROEs) were non-existent. Very little information came down to us, and when it did, there was considerable disorganization, miscommunication and even conflicting orders. The one thing everyone was sure of was that no one knew what was going on. Every attempt was made by the command infrastructure, however, to keep everyone in the know. Simple steps such as routine orders were being published, shifts and schedules were being created and camp routine was being established. The standard direction that a soldier expects, such as where to wash, where and when to eat, and travel restrictions, were laid down. As far as normal routine went, we were set. What was missing, though, was the "big picture."

In this instance, it is very easy to blame everyone from the commander of the battle group (BG) all the way down to the section commanders. Working in the headquarters (HQ), however, it was clear that information from higher command (national, international, UN) was sporadic at best. Looking from the ground up, there was no clear mandate, no direction and no clear mission. Everybody was doing "his or her thing," but it seemed that there was no greater mission or overall objective. In the first couple of weeks, there was a considerable amount of "hurry up and wait" and "we will let you know when we know." In other words, we had a battalion of troops with neither direction nor purposeful activity. Having no clear direction or mandate makes soldiers nervous. There was considerable confusion down the line as to our overall mission and objectives. Were we to disarm these people? What happens if they start shooting (which happened often, especially at celebrations and weddings)? How far do we go? What are we authorized to do?

This unfortunate situation, however, was out of everyone's control. One only need review the UN Resolutions from late 1991 to the summer of 1992 to see what went wrong. At every turn, the situation was frustrated at the highest levels by the parties who would agree to a resolution and then violate it, sometimes within hours. As fast as information about new Resolutions and "possible" mandate changes was produced, disseminated, and put into place, the face of the war had already changed. Constant cease-fire violations and persistent troop movements on all sides made the situation on the ground tenuous at best. And make no mistake, this was a war and we were in the middle of it. All parties were content to fight it out, whether the UN was in the way or not. Commanders from the highest levels all the way down to the individual sections had their hands tied by either too much or not enough information. By the time information flowed down to section commanders, it was stale. It must be understood that the "Blue Helmet" meant nothing to the combatants. There was no respect for the UN banner, the UN troops or any other outside agencies. This appeared to be very unfamiliar territory for all involved.

Most soldiers hit the ground with a "block of trust." Soldiers expect clear direction and ROEs. This mentally protects the soldier from having to make decisions where his or her own life, or the lives of others, may depend on one's own actions. There is no detail that is too small to include, whether it be about returning fire or camp discipline and routine. Not having these basic instructions chips away at the block of trust. Even "no news" is news. Keeping everyone informed, even if the answers are not available, builds on the trust block instead of wearing it away.

TRUST IN THE TROOPS

The scene around Daruvar, Croatia where the Canadian contingent or CANBAT was initially harboured was an uncomfortable place. Everybody, including the civilian population, had a gun. Side arms, long rifles, grenades and combat clothing in various states of completeness (mixed with bravado and testosterone) were clearly evident. People with guns and a large amount of alcohol do not make for a comfortable situation. To further complicate this problem, our initial camp was set up in a soccer field directly across from the largest bar and hotel in the area. At any given time, clerks, cooks, supply technicians and infantry alike would be in the same rest area (the local hotel/bar) as the host nation's military, para-military and ad-hoc fighters, the general population and the staff of the hotel/bar.

The presence of other UN forces compounded the tension between the local population and the Canadian contingent. For the most part, Canada is a very disciplined nation when deployed. Command gives instructions, rules and boundaries for acceptable behaviour, and in a very wide sweeping way, commanders pass trust through the chain of command all the way down to the private soldier. I think it is safe to say that the trust that is granted to Canadian troops to use their own self-discipline and control is well placed. Canadian soldiers are intelligent and trustworthy. Often during this mission, you could observe the troops from other nations acting unprofessionally: getting drunk, becoming disorderly and looking for a fight. Their comrades, instead of diffusing the situation and protecting each other, would encourage such behaviour. Herein lies the problem: the civilian population and combatants do not "see" Canadian soldiers, British soldiers, Jordanian solders and so on, rather they see UN soldiers. They do not differentiate between the participating countries, and on this mission in particular, we were viewed as unwelcome troops and potential "invaders." It is hard enough for Canadian commanders to gain understanding from the host country when they must continually remind other commanders from other countries to monitor the actions and activities of their own soldiers. From the perspective of a command hierarchy, it must have been frustrating, particularly in the "political forum" in which we were operating. It could not have been easy for the BG commander or the BG regimental sergeant-major (RSM) to gently and eloquently persuade commanders of other nations to control their troops.

While working one-on-one with troops from other nations, it is easy to see where they fail. Most foreign nations have no trust in the "the common soldier," nor do they have trust in their junior and senior non-commissioned members (NCMs). First-line decisions are entrusted to the officer corps and only the letter of the law is acceptable to them. In most cases, discipline is swift and harsh. Insight and conversation is not sought from the NCM corps, and these soldiers often make their decisions based on fear of reprisal rather than common sense. The trust and confidence that command has and continues to provide to Canadian troops to think and act independently is normally not afforded to the soldiers of other countries. Not having this dynamic of trust and confidence in one's subordinates creates a situation where they are not practiced in making their own decisions; without this trust and confidence, they will inevitably fail when pressed to make the right decisions.

In fact, this philosophy is rooted in the very core of the Canadian military. One need look no further than the second stanza of the Royal Military College of Canada's motto: "DUTY – Do what you ought to do when you ought to do it, whether you like it or not, and without having to be told or watched."[2] You would be hard pressed to find Canadian commanders who do not have a high level of confidence and trust in their troops to "do the right thing." In fact, when Canadian soldiers get themselves into a bad situation, anger (the most primitive and immediate reaction of a superior), is in most cases quickly replaced with profound disappointment owing to the violation of trust that has just occurred.

Trust your soldiers to do the right thing. Prepare them and exercise their minds before they get to where they are going. Knowledge is power, but it is nothing without the opportunity to practice it. Teach your troops the right way to do things, be clear on your expectations in all situations, and speak to them with respect. Give them opportunities to grow and chances to fail in a safe environment. Follow these principles and that soldier may someday keep you from harm. They will not break your trust and you will not feel that "thud" of disappointment if things do go wrong.

TRUST IN THE WORD

Author Douglas Adams (1951-2001) said in his novel *Mostly Harmless*, "Nothing travels faster than the speed of light with the possible exception of bad news, which obeys its own special laws."[3] We would be well advised to remember this in any situation that confronts us as leaders. It also seems to be likely that the speed at which bad news travels is directly proportional to the amount of (usually lack of) good information that is available. No one needs lessons on how stories and "facts" get embellished and expanded upon. These types of situations can be avoided, but be sure of this, you cannot eliminate them. What is important to recognize is that the troops need to have confidence in their leaders. They need to trust that they are being given good information. It is incumbent on leaders at all levels, not only those who are in theatre, but also those in the rear party, to quickly and effectively stop rumours, clarify issues and quickly and decisively deliver as much information as possible.

During Op Harmony, there was an extremely high amount of communication back home as most of the BG had come from Lahr or Baden, Germany. This was a unique situation where families were only a one-day drive from theatre. In today's operations, it is common to have communication blackouts when a significant incident occurs (which in themselves create a high level of tension both in theatre and back home). Communication blackouts were almost impossible to achieve since we were so close to home. The information network from theatre back to families was fast and intricate, and although there is no real evidence, it often seemed that the families back in Lahr and Baden knew things before command did. Whether the information was reliable, embellished or completely unsubstantiated, word travelled fast and furious. It is important to recognize this as families become more anxious and worried as their loved ones speak to them about the mission. Even worse was when families told us information that had been written in newspapers or seen on television. With families back in Lahr and Baden, news was available from Sky Satellite services and the local German news broadcasts. These services would often report on the entire war and broadcast very vivid and graphic news. It is naive to think that this can be stopped.

In the macro perspective, it is easy to dismiss "stop the rumour" orders, but as the command infrastructure closes in on the company, platoon and section, crushing the rumour mill becomes increasingly difficult. Operational rumours are for the most part accepted as just that, operational rumours. These types of rumours can often be squashed or more importantly clarified at the micro level with constant briefings and good communications. It is incumbent upon all levels of leadership to remain vigilant in listening for, and acting upon, unsubstantiated rumours and to "put them to bed" with great resolve. This is a huge undertaking and one that is easy to get trapped in. In no time, and without being aware of it, you can become part of the rumour mill yourself.

Even more damaging though are the rumours to and from home. It seems that every mission has the same rumours: "somebody's wife is doing this…;" "somebody's husband didn't even wait a day before…;" "someone told someone that someone in "Charlie" Company is visiting somebody in "Alpha" Company…;" "someone called and told their wife/husband…;" and on and on it goes. Often, they are just stories that float around camp that become increasingly inflated and embellished. Innocent connections or misunderstandings can be quickly blown out of proportion and these rumours will inevitably make it back home. There is very rarely any solid evidence of indiscretion and often these incidents take the focus off of the mission. The soldiers in the trench beside you need to have their head in the game and to not be flooded with complications derived from rumour and innuendo. Leaders, specifically down to the section commander, need to be aware of all of the personal lives of their troops in order to avert potential situations like this.

Trust, one-on-one trust, is the only thing that will prevent these types of situations from festering into an uncontrollable mess. This trust must go both ways, and if possible, extend out to the families of your subordinates. On missions like this, faith and trust put in leaders must be rock solid so family members are spared undo hardship and anxiety. What is being spoken of is not "a code of silence." In fact, it is exactly the opposite. Soldiers need to be open and frank with their leaders, and leaders must recognize and respect the trust that the soldier is placing in them. A violation of this trust may become insurmountable for section commanders. Violation of the private lives of their troops will ensure that neither their subordinates nor their superiors will ever trust them again.

The residual effect of course is confidence on the part of all involved. It is important to remember that families back home are just as much a part of the mission as the troops on the ground. They need their information, their network, their faith and their confidence. The average soldier has enough to worry about on a daily basis without having to worry about what his family is "hearing" back home. By creating a package of trust, trust will be built between families and the unit, soldiers and their leaders, and there will be a lot less anxiety and frustration on the part of everyone.

BREAKING TRUST FROM WITHIN

> One other part of leadership is your concern and efforts on behalf of the welfare of those who serve under you. You must prove beyond doubt that you are fair and just in your dealings with them and that you genuinely like and respect them. In all circumstances and regardless of cost to you yourself, you must place their well being ahead of your own.
>
> ~ Air Commodore Len Birchall.[4]

Trust is generally viewed in both the macro and the micro simultaneously. Trust in the greater whole is often met with a degree of scepticism from the onset, and when that trust is violated, it is normally not a surprise or really unexpected. The cut is not so deep. When trust, however, is violated at a micro level, the effects compound at a rapid rate.

In the first few weeks of the mission, the orderly room (OR) was working 18-hour days, sometimes longer. The important part to remember is that we were the first on the ground. There was no administrative infrastructure and everything was being set up from scratch. On top of our daily tasks of logging the ever increasing amount of message traffic, creating nominal rolls, providing identification services, and all the other things that a HQ OR has to deal with, the junior NCMs were also required to undertake security duties.

As the OR staff looked around, our administrative counterparts in other sections (operations and company clerks, for instance), looked clean, rested and were conspicuously absent from security shifts. Although this was brought to the attention of the supervisors, long office hours combined with the various responsibilities associated with security went on for weeks. Meals and down time were also ignored by supervisors who insisted on tight deadlines and the completion of items, despite the welfare of their subordinates. Intervention from outside of our administrative group eventually rectified this difficult situation, however the damage had been done. The level of trust between subordinates and supervisors dropped dramatically and remained extremely low throughout the entire mission.

What was missing was a lack of acknowledgement from the supervisors. There was no recognition of the amount of time and energy being put into the work. Simple interactions such as asking subordinates if they had eaten or had been provided some rest time would have made a world of difference. Instead, the general lack of interest in the welfare of the troops gave the impression that the leaders of that organization cared only for their own well-being and comfort. The impression was: "they never miss a meal, they had a shower, they got a full night's sleep." To be fair, that statement will never be proved or disproved however, and we must remember that we are dealing with the impressions of superior activities. It is well known that the "impression" of wrongdoing holds greater credence than the truth if not acted upon quickly and decisively. Once the foundation of wrongdoing (factual or not) is laid, you will never get the full trust of your subordinates. There will always be suspicion and distrust whenever decisions are made, orders are issued or instructions are given. It breeds contempt and is a constant distraction to the individuals who feel that they have been ignored or forgotten. Regaining the trust of those subordinates is highly unlikely, ever. To this day, I still believe that one individual rode the backs of his subordinates all the way to the top and this individual would never recognize his wrongdoing, his lack of acknowledgement of his subordinates, or any other transgressions. I have some faith that he was and probably still is oblivious to this and can only hope that someone has pointed out how his actions demoralized his subordinates. The interesting part is that as you read this, you will know exactly who I am writing about. We have all met people like this and work with them every day. Even if you do not have direct contact with a leader like this, you most certainly have heard of one, or know of one who acts the part.

I also had my roll to play in this. My failure in this relationship was not having the courage to calmly bring it up and address the situation. My impression of this person may have been completely wrong, but by not addressing it, my feelings about this trust violation remain to this day.

Despite my feelings, however, I did walk away with some valuable insights and lessons learned. I know how to monitor both my subordinates and myself. I am keenly aware of the needs of my subordinates and do not allow things to fester. Issues are brought out into the open and resolved. My subordinates and my supervisors are made aware of any pressures or obligations that may cause me to seem to be taking advantage of them or the system. It is important to know how to effectively deliver this information as well. There is a fine perceptual line between informing the staff of what is going on and "whining" about all the pressures that you are under. Delivery is key. If you have done it correctly, offers of support or assistance will be forthcoming, and at a minimum, you will be able to tell by their faces that they understand. These simple actions do not make a leader appear weak or soft. In fact, it is inspirational to the troops. It creates a bond of trust and makes subordinates feel acknowledged and cared for. We are taught in the military that our leaders will ensure to the best of their ability that the welfare of their troops will come first. As a supervisor, should you choose to ignore this simple leadership principle, the trust your subordinates have in you will dissipate quickly. It will create tension and place leaders in a position where they either have to justify their actions or be seen as one that only cares for themselves and pleasing their bosses at the cost of their subordinates.

TRUST GIVEN, TRUST ACCEPTED

Security of the HQ area was eventually taken over by a section of infantry (from a larger company) of The Royal Canadian Regiment (RCR) that was attached to the BG. These soldiers were led by a supremely confident individual who had a genuinely caring attitude that showed in the attitudes and professionalism of his soldiers. During a late evening in the office, one of this sergeant's soldiers was providing security on the main floor of our building. It was a cold and dark evening with a biting wind. I was taking a small break and a stretch and approached the young private. I asked him if he would like a coffee, if he needed to take a break for 15 minutes, or if he needed to go to the bathroom. I mentioned that I would be happy to take his place. He told me that his sergeant would not like that and that he would be disappointed in him if he was not here when he made his rounds. Although the attitude of this young private is difficult to portray on paper, his statements were made not because he feared getting in trouble with his supervisor, but rather because he feared breaking the bond of trust and respect that existed between the two of them.

The next evening I was privileged enough to meet the sergeant who had inspired this young private. He was not what I expected at all. Soft spoken and the true look of an honest man. I spoke to him about the events of the night before and relayed to him my offer that I had made to the young private. He explained to me that the private had told him about our conversation and offer. He went on to explain that "his troops" followed orders to the letter, and even though he would not have been upset to find me taking the place of one of his men for a few minutes, he was proud that his soldier had stuck to his post. We agreed that night that any evening when I was working nights and his troops were on security, it would be acceptable for me to take over for a short duration while they warmed up, had a coffee, and so on.

Over two decades of service, I have interacted with people at all levels of rank and authority. So why does this one particular instance, two days out of a 180-day mission, remain so clear? What is it about this small interaction that has managed to permanently implant itself in the foreground of my memory? There were no heroic acts, no great decisions, and no life-threatening or life-saving events. It is truly the bond of trust. There is no particular reason that I should have been included in the web of trust between the sergeant and his troops, and yet, there I was. Through honesty and conversation I was brought into the group; this came at a cost, however, the deal being that I would have to stand most of a shift with him on my down time and learn the ropes. This is a true relationship. Trust granted, trust offered, trust earned, trust accepted and trust exchanged. There was a willingness on the part of all parties to engage in a relationship of trust and respect. There were no spoken words of trust, rather it was implied. In many of these situations, the unspoken word has great weight and great significance. There is no fear of punishment or reprisal, nor is there any fear of disciplinary action. The one great fear in this sort of relationship is that you will let down the one who has granted you trust. The proper ending to this story would be a great relationship and friendship that this sergeant and I shared over the years, but that is not the case. It was a short interaction that had profound meaning to me and brought home the meaning of good leadership, morale, and more important, how trust is a vital tool, not only in theatre, but in a static role as well.

TRUST IN ONESELF

The C-7 rifle is a gas operated, air cooled, magazine fed, semi-automatic or automatic 5.56 mm weapon, with 6 groove right hand rifling. Its magazine holds 30 rounds and has a cyclic rate of fire of 700-940 rounds per minute. It weighs 4.4 kilograms with the C-79 optic sight and a full magazine. It can be fitted with a bayonet for close-quarter fighting. It is capable of placing effective fire on a target at approximately 300 metres. It is your personal weapon.

So there we are. After 20 years and even after being retired, my training has provided me with the tools to understand what my personal weapon is. Countless drills and dry training prepared me to handle the weapon safely. It prepared me to aim the weapon effectively. Training gave me the confidence that even in the most difficult times, I could rely on my weapon to provide protection. I was confident that my weapon worked. I trusted the weapon … an extension of my body to do exactly what it was designed to do … put holes in "targets." As the mission progressed though, I became less confident in my weapon. Not so much its mechanics, design or reliability, but rather the thing just behind the butt of the rifle … me.

In a chapter entitled "Psychological Effects of Combat," authors Dave Grossman and Bruce K. Siddle open the segment entitled "Overcoming the Resistance to Killing" with the following paragraph:

> By 1946, the US Army had accepted [S.L.A.] Marshall's conclusions, and the Human Resources Research Office of the US Army subsequently pioneered a revolution in combat training that eventually replaced firing at bull's-eye targets with deeply ingrained "conditioning" using realistic, man-shaped, pop-up targets that fall when hit. Psychologists know that this kind of powerful operant conditioning is the only technique that will reliably influence the primitive, midbrain processing of a frightened human being, just as fire drills condition terrified school children to respond properly during a fire, and repetitious, "stimulus-response" conditioning in flight simulators enables frightened pilots to respond reflexively to emergency situations.[5]

After reading this, those of you who went through Canada's cadet movement will suddenly realize the significance of this statement. Even during my time in the cadet program, we studied and practiced marksmanship. To this day, the cadet movement still uses "bulls-eye" targets.

Soldiers today fire at the figure 11 or figure 12 human silhouette-style targets. These are two-dimensional human shaped figures with angry facial expressions pictured in such a way that they appear to be running towards you. I showed a picture of these targets to some of my civilian colleagues that I now work with and asked them if anything struck them as being out of place or seemed odd. Universally and within seconds, each person I asked identified the same thing: "They have no eyes."

So how was I going to react to a target with eyes? This is a situation that cannot be resolved with training. The act of shooting another human being or even the prospect of it is disconcerting for most people and I would submit even more so for the soldier. Soldiers put themselves in scenarios where they may be required to fire their weapons. I would suspect that most soldiers do not confront the scenarios under 25 metres and I would probably tack on an additional 25 metres to that too. The ability of a soldier to use deadly force diminishes exponentially as the distance between himself and the target shortens. On this mission, the vast majority of interactions tended to be up-close. These were at checkpoints and observation posts. Tension and miscommunication often led to situations where even routine travel could create a dangerous situation.

During this mission, I was put to the test on more than one occasion. During a crossing of the Sava River bridge *en route* with a French battalion to Sarajevo, we were confronted by a large contingent of Yugoslavian National Army (JNA) troops holding one end of the crossing. The situation became very tense in a very short period of time. The greater acts of these moments seem to fade, but what I most remember is the utter fear that I felt at not being in control of the situation. Weighing heavily on my mind was the fact that we were in a convoy led and commanded by another country. As I recall, there were only six Canadians, the highest rank being a warrant officer. It is unnerving to have the control of a situation in the hands of a commander from another country. When under the command of Canadian commanders, it is easy to predict their actions. Every effort will be made for a non-confrontational resolution. The chances of having to engage will be minimized and every effort will be taken to keep everyone on both sides of the situation safe. Being led by a commander from another nation created considerable doubt in my mind and this fact added additional anxiety to the situation. My level of trust towards the unknown French commander was very low.

As we had spent considerable time in our vehicles, it was decided that we should dismount and await the arrival of the local commander of the JNA. The order was "no weapons," but to leave the doors of the vehicles open so as to provide quick access to them in case they were needed. The sergeant that I shared a vehicle with carefully removed our weapons from the racks in the vehicle and laid them across our seats. We pulled our jackets over our side arms so they were out of sight and then dismounted. I knew full well that if anything should erupt suddenly the chances of me reaching my weapon were slim to none.

Once out of the vehicle, I was immediately met with an AK-47 at my chest with a very stern looking man on the other end. A few shouts from up the road and the man relaxed and took his weapon off me. No amount of space in this book could cover the amount of traffic that went through my brain in that very brief encounter. At this point, I would like to share one of the most valuable pieces of information any soldier can know: never underestimate the power of cigarettes. Even if you don't smoke, carry them!

I am sure that I was trembling at the prospect of being shot squarely in the chest, but I somehow managed to make the international sign of two fingers to the lips indicating that I would be fishing through my uniform for cigarettes. For the first time, my new best friend shared a smile with me and he started rummaging through his clothes. We swapped a smoke and through some broken attempts we discovered that we both spoke some very bad German. Both struggling to translate in our heads from our native languages, I think we came to an agreement that we were both very tired and we both thought this was crazy. I was later informed by one of the French officers that my calm actions and my "befriending" of the man who had put a loaded weapon to my chest did not go unnoticed by the negotiators farther up the road. Our interaction had served to help diffuse and calm down the situation. Heroic? Not even remotely. Lucky? Absolutely!

The important point here is recognizing my underlying actions. There was no training that could have prepared me for that situation. The situation could have deteriorated quickly if I had carried my weapon. Without thinking, my mind took me to the primal area of self-preservation. There were no concise thought patterns that took me to the point of offering a cigarette to the man; rather, it was a "natural" reaction. Anyone who smokes can tell you that after a stressful moment, your hand reaches for the pack. The only presence of mind that I had was to show the man across from me what I was going to do when reaching inside my clothing. The simple act of sharing with this man created a degree of trust between two people that had no reason to trust one another.

What really needs attention though is what happened after we got through the checkpoint. I was flooded with a horrible case of the "what ifs." I replayed the scenario over and over again in my head with every possible outcome. I studied my actions and reactions and all of the possible ways that it could have went wrong. Over a decade later, my heart rate still increases when thinking about it and that shiver comes back. The only conclusion I had then and still come to today is that the trust I had in myself, to deal with such a situation, had been tested and I had passed. Not all of us had the same experience though.

Every Canadian had been tested several times throughout that time on the road to Sarajevo. We would endure many situations with our French commander over the days that we travelled with the convoy. My personal trust and respect grew with every passing moment and every situation. On one occasion, our convoy arrived at the airport in Sarajevo, and unfortunately, one of our small contingent of six needed to be medically evacuated. This individual held it together for as long as he could, but relented once the destination was reached. Who it was is not important. What is important is the recognition of the block of trust that we had in ourselves. It is easy for situations to chip and chisel away at that block of trust until it splits. Everyone's block is a different size and a different shape. Experiences will either break chunks off or build on it.

Once at the airport, routine was often interrupted by gun, mortar and rocket fire. There was a bunker on the bottom floor of the building that we occupied when things got really bad. On the occasions (which were usually daily) that we would come under fire, we would head to the bunker. Good practice when moving everyone to a bunker is to ensure that guards are posted at entry points and areas of concern falling back all the way to the bunker. I always volunteered to act as one of these sentries. This was no act of bravery by any stretch; rather it was an act of self-preservation. Now it may seem odd that I would rather be out in the open than in the relative safety of a bunker, but I fear confined spaces and have a fear of being buried alive. My block of trust in myself tells me that I will perform better out in the open than inside a bunker.

There were many days and hours spent guarding the bunker or various parts of the building during shelling or other actions. So much so that my continued presence outside caught the attention of one master warrant officer. I always saw him making the rounds and he always saw me. On one particular day, he ventured past the formal conversation of how much ammo I was carrying, what my arcs of fire were, and if I knew where our other sentries were and where to fall back to if required. He commented to me that he always sees me on sentry duty and asked rather bluntly if there was some sort of "command" issue he needed to deal with. Hoping that my reasoning made sense to him, I told him of my discomfort of being in the bunker and felt that from my own perspective, my reactions to situations would be more "rational out here than in there." With a little Mona Lisa-like smile he said two simple words, "roger that," and went on his way. The tone of his voice cannot be expressed here, but I got the impression that the reason we saw each other out in the open so much is that perhaps we had the same fear and twisted logic. My personal block of trust grew that day. I knew I was not alone in my fear and that somebody trusted that my judgment of myself was worthy and rational.

THE FINAL INSULT

The "regimental family" is one thing that you can believe in. They will support you and your family and they will be there for you. Everyone looks after everyone. Prior to deployment, I was posted to 4 Svc Bn in Lahr. Remembering that this was at the time of the base closures, the battalion had been considerably downsized to less than company strength. There were, however, still command elements and staff working in the battalion. As the end of the rotation neared, groups were heading back to Lahr and Baden. I returned with a small group that left from the Zagreb airport and flew directly to the airfield in Lahr. On this trip, the main body of about 50 was made up of troops from 1 R22ndR, with three soldiers from "November" Company, The Royal Canadian Regiment (in Baden), myself, and one other support staff from 4 Svc Bn.

Nothing is more exciting for a soldier than coming home: being with family, friends, your own bed, flush toilets and all the "creature comforts." You get to eat when you want, sleep when you want, have a shower when you want and relax the way you want. There is no sentry duty, no rifle, no helmet and no flak vest. It is a great moment to come home. As the aircraft taxied in at Lahr, it was clear that there were many families waiting for us. The two of us from 4 Svc Bn were single, so there were no wives or children awaiting our return. The rear party commanding officer (CO) of 1 R22ndR, in addition to the CO of 3 RCR and his RSM, were also present. I remember thinking to myself as we deplaned: What an incredible display of appreciation on the part of the RCR! Three of their soldiers were coming home and the CO and RSM made a point of making the hour long drive (on a weekend as I recall), just to meet them.

After getting off the plane, we went through the repatriation process: quick interviews, shots, blood tests, leave passes, etc. At the end of the line were a couple of tables that had various items on them. Some departing gifts as it were: a plaque with my name on it, a Swiss Army-style folding pocket knife with my name and the mission engraved on it and some other goodies that I cannot remember. The battalion was aware that I was coming home and they made every effort to get me through the process quickly and efficiently so that I could go home to Canada while on leave. It was the "thought" that made me feel so much a part of the mission even as it came to an end.

With departing gifts, plane tickets and leave pass in hand, I was ready to head home to my apartment just outside of Lahr. I walked out the back door of the hangar and suddenly realized … I had no way to get there. The RCR had sent a bus for the families and soldiers to take back to Baden. The others from 1 R22ndR all had family or regimental drivers to take them home. I popped back into the hangar to ask if there had been any transport laid on for us and I was informed that all the units had been advised of who was coming home on what days and to make sure that they had transport available. It quickly became evident that the two of us from 4 Svc Bn were not getting a ride home. I also realized that I did not have any local currency to get a taxi with. I had received my close-out cheque from the mission and had money in the bank, but this was 1992 and cash machines were not available in the quantities that they are today. The two of us sat there with no money, no transportation and no way home. One of the staff from the hangar called the duty officer at 4 Svc Bn and was informed that they had no vehicles or drivers to help out. After a couple of hours of trying to get us a ride, someone from 1 R22ndR came in a vehicle, picked us up and drove us home.

I still had a day or so before my flight left for Canada and I went over to the store to pick up some gifts and to do a little shopping. While I was there, I bumped into the now-commanding officer of what was left of 4 Svc Bn. Oddly enough, he had been my company commander before I left for Yugoslavia. He greeted me and said that he had heard that my colleague and I were returning this weekend and that he was glad to see me. I thanked him and mentioned that we could have really used a ride home. Standing beside him was his wife (presumably), who before he could reply to my comment said to him, "You knew they were coming home, and you didn't go and meet them!" She then turned away with her cart and walked away. Without another word to me he turned and went after her. Figuring that since I was posted to a new unit in Baden after my leave I would never have to see him again, and the punishment that was going to be doled out by his wife was far better than any rant or guilt trip that I could lay on him, I was satisfied that my point was made. The trust in the regimental family in this instance was completely violated. One is left to wonder what it would have taken for this individual to accept the responsibility of meeting his soldiers coming home from the most brutal UN action since the Korean War.

THE LAST WORD

At the beginning of this chapter, I mentioned about learning from good leaders and from bad ones. It is my hope that I have shown the many aspects of trust and how easily it can be gained or violated: how trust is given, received, passed and accepted. Specific individual names were not used in writing this since it is not important or, I feel, appropriate to single out individuals for what my perceptions of their shortcomings were. More important is how you place yourself within these stories and how you react to them. It is important that we learn from the actions or inactions of others. I hope that my experiences create the same type of "mental checklist" for you when making decisions that will have an impact on those that you lead.

In closing, going through this exercise made me realize many things. Respect and trust with a highly evolved awareness of common sense are three ingredients to being a good leader. None of these individual items can stand alone. Trust and respect must be bi-directional. It must go up and down the chain of command to be effective. Using common sense and thinking about the effects of your decisions, weighing options and giving and taking advice from superiors, subordinates and peers will earn their trust. With that trust in hand, you will earn the respect of all around you. Loose that respect or trust and you may never regain it.

ENDNOTES

1 Katherine Barber, ed., *The Canadian Oxford Dictionary* (Don Mills: Oxford University Press, 1998), 1558.

2 RMC's motto reads: Truth, Duty, Valour.

3 Douglas Adams, *Mostly Harmless* (New York: Ballantine Books, 1993), 1.

4 Excerpt from a speech on leadership given by the late Air Commodore Len Birchall. See Canada's Air Force News Room, Holly Bridges, 6 April 2006, http://www.airforce.forces.gc.ca/site/newsroom/news_e.asp?id=1513.

5 Dave Grossman and Bruce K. Siddle, "Psychological Effects of Combat," in Lester Kurtz and Jennifer Turpin, eds., *Encyclopedia of Violence, Peace and Conflict*, Vol. III (San Diego: Academic Press, 2000), 146.

CHAPTER 6

OPERATIONAL CHAPLAINS: ESTABLISHING TRUST WITH THE RELIGIOUS OTHER THROUGH THE BUILDING OF RELATION

Major S.K. Moore

Theologian Miroslav Volf enjoins Western thinkers that even though religious wars are no longer a threat, we should not underplay the existence of a religious component in much of contemporary conflict.[1] R. Scott Appleby nuances Volf's comments further in stating, "In certain settings leaders appeal to religious identity as a means of exploiting ethnic and tribal animosities; in other settings they invoke religion as a means of transcending differences and unifying rival tribes."[2] Arguably, more than at any other time in recent history, religious fervour is stoking the embers of conflict around the globe and, as such, is gaining serious attention in many quarters.

The dynamic of trust and how it may be cultivated between military contingents and local ethno-religious communities in the grips of open conflict or its residual effects forms the focus of this chapter. The role of operational chaplains as religious leaders in their own right, and their unique capacity to build relation with the religious *other* thus engendering trust among conflicting faith group communities, will also be examined. The challenge for the reader, therefore, will be to envision such trust-building among local religious leaders as an added dimension of operational mandates. Salutary to this notion is the recent surge of interest in the religious domain and its potential contribution to the resolution of conflict. This has become poignantly relevant in areas of the world yet convulsing from the influence of more aberrant expressions of religion.

In July 2007, the Washington-based Center for Strategic and International Studies (CSIS) released a report entitled *Mixed Blessings: US Government Engagement with Religion in Conflict-Prone Settings*. At the heart of this landmark document is the acknowledgement that: 1) successful resolution to conflict abroad will necessitate the engaging of the religious *other* and, 2) at present, government departments and agencies are inadequately prepared to undertake such endeavours. Continuing further, the authors state:

> In a world heavily influenced by religion, U.S. government intelligence, military, diplomatic, and development tools must be

> properly prepared to engage these religious elements ... Religious leaders, organizations, institutions and communities can mobilize religion to sanction violence, draw on religion to resolve conflicts, or invoke religion to provide humanitarian and development aid. To engage successfully, government analyses, policy, training, and programming must fully incorporate an understanding of the varied roles for religion in conflict-prone settings.[3]

The recommendations issuing from this report call for a sweeping increase to the knowledge base concerning the religious dynamics of today's world. Among these are: additional government expertise in all departments with respect to religion; expanding outside partnerships with academics in related fields; and, linking this community of experts as a force multiplier. Poignant to the discussion to follow, and under the heading "Sensitize Programming to Religious Realities," CSIS forcefully advocates the engaging of a broader range of leaders in conflict-prone parts of the world, specifically identifying religious leaders and faith-based groups as potential partners integral to resolving contemporary conflict where the manipulation of religious fervour has become a factor. The authors explain:

> Programs should seek a fuller range of religious representatives abroad and engage with less traditional – and possibly less welcoming – religious leaders and audiences, recognizing not only 'religious moderates' but also 'religious conservatives' as opinion leaders and possible drivers of change ... government should enable increased partnerships with previously excluded faith-based groups abroad, actively pursuing them as partners in facilitating inter- and intra-faith dialogue.[4]

More relevant still to the discussion to follow is the report's citing of the strategic role of operational chaplains. "As official representatives of religious denominations, chaplains are the de facto experts in religion on the battlefield."[5] Again, the report states, "Of all military personnel, military chaplains are best suited for religious liaison work, according to a report from the U.S. Army's Peacekeeping and Stability Operations Institute."[6] More than any other contingent member, the chaplain embodies a natural and transcendent bond with their local counterparts in theatres of operation. As such, they offer inroads to local communities where the intermeshing of religion and culture blur any lines of distinction.

The theme of this iteration of *In Harm's Way* is that of trust. I have purposely drawn on the CSIS report as a prelude to this discussion to underscore the shift in strategic thinking that is underway with respect to the implications of the religious *other* in the resolution of conflict globally. A word of caution must be offered at this juncture: engaging the religious *other*, moderate or conservative, welcoming or hesitant, relies heavily on the building of relation and the inherent trust that that process generates. To reduce such relation to a technique undermines its integrity, an integral aspect of engendered trust and, as such, must be resisted. Throughout this chapter, the building of relation and trust will be seen to be related terms as the latter emerges from the former: no trust without relation, no relation without trust.

Time and again, operational chaplains have demonstrated their ability to bridge relationally to seeming distant local communities by way of their religious leaders, "[individuals] who may be well recognized and respected within that network or geographical region."[7] Where the desire exists within the religious *other* for a reconciling of differences and a return to coexistence/relation, movement toward such interdependence may emerge. In instances where entrenchment is such that enmity and bitterness prohibit communication, it is conceivable that a chaplain's engaging of local clerics may model inclusivity, generating an openness to dialogue among estranged ethno-religious faith group leaders. It is journeying in relation with the religious *other* that such trust is nurtured and established. In an environment of earned trust, a new vision of the *other* may begin to emerge. The effectual engaging of moderate local religious leaders by a chaplain may in turn create similar possibilities with those who are on the fringes of opposing groups. In time, and in concurrence with command, we may learn that there are other ways of "engaging the enemy."

The following discussion is based on actual events that occurred in the Bosnian theatre of operations. Brought forward will be the benefits to mission mandates of building trust among the religious leadership of ethno-religious communities in conflict. In this instance, the combined efforts of chaplains from consecutive rotations aided conflicting groups in capturing a renewed vision of the humanity of the *other*. Illuminated also will be how, with the support of command, such engendered trust led to an unprecedented occasion of interfaith celebration among estranged faith group communities in the midst of overt conflict. Such establishing of trust through the building of relation gave rise to unparalleled cooperation among these faith group leaders, symbolically signalling to their respective ethno-religious communities that the future lay not in fighting but in fraternity. Operational ministry in this vein may be best articulated as the seeding of reconciliation due to its embryonic nature.

UNPROFOR, OPERATION CAVALIER, CANBAT II, 2 RCR BATTLE GROUP

Visoko & Kiseljak, Bosnia-Herzegovina, Rotation 0, February-May 1993

In the spring of 1993, I deployed with the 2nd Battalion, The Royal Canadian Regiment Battle Group (2 RCR BG) from southern Croatia to the outskirts of Sarajevo, a city very much at war. The main body of the BG resided in the town of Visoko with more than a company in Kiseljak, both communities on the edge of the capital. Ours was Rotation 0, with an area of responsibility encompassing representative populations of all three ethno-religious groups: Serbian Orthodox, Croat Roman Catholic and Bosniak Muslim.[8] At the outbreak of hostilities, the majority of Serbs left the town for safer environs, leaving it to the Croats and Bosniaks.

Of importance to this discussion are the beginnings of relation that Padre Mari Eugenio (Roman Catholic priest) and I established with the local religious leaders of the Roman Catholic and Muslim faith communities. These early developments provided the footing for further dialogue with the chaplains of the soon-to-deploy 2nd Battalion, Royal 22nd Regiment (2 R22ndR) from Valcartier, Québec.

Upon learning that the newly arrived Canadian contingent to Visoko had with them two Christian "priests," an invitation to dialogue was immediately issued from Imam Asim Azdahic, a cleric responsible for the supervision of 60 mosques in the greater Visoko area. In the Imam Association conference room of their downtown offices, five senior imams conversed with us over refreshments for an entire afternoon. Muslim hospitality was extended with openness, accompanied by a tour of the famed "White Mosque" of Visoko. The imams were inquisitive of life in the West, in particular, the fervency of Christian belief and praxis.[9]

During the remaining months of the rotation, greater trust began to develop between us and both the Franciscan and Muslim communities. Unfortunately, hostilities were such that overtures toward the Serbian Orthodox were not possible at this time. Regular visits to both Roman Catholic and Muslim faith communities were, however, complemented with the sharing of humanitarian assistance as it became available from Canada via convoy from the port of Plocé on the Dalmatian coast of Croatia. As always, care was taken to be impartial. *Merhemet* (Red Crescent) was particularly appreciative of the clothing and non-perishable food items. Hospitality was extended to the leadership of both faith communities to visit the CANBAT II compound to meet the commanding officer, to take a tour of the contingent chapel, and to share in a meal among the troops. Imam Azdahic graciously accepted and invited us to his own home to meet his family and for refreshments.

As visits became more frequent, the warming of relations became evident. An emerging trust led Imam Azdahic to share openly his intense longing to return to more amiable times when Bosniak, Croat and Serb lived as neighbours. This genuine expression of the "will to embrace"[10] the religious *other* bore witness to his nascent desire for reconciliation among these ethno-religious communities still in conflict. His parting words to me prior to my redeployment to Canada were to tell the West of their story, that they were a peace-loving people who did not want to be at war with their neighbours. His was a moderate voice courageously calling for reason where there was chaos, an opening of the *self* to the religious *other*, where there had only been exclusion, isolation and alienation.

UNPROFOR, OPERATION CAVALIER, CANBAT II, 2 R22ndR BATTLE GROUP

Visoko, Kiseljak & Srebrenica, Bosnia-Herzegovina, Rotation 1, May-October 1993

The subsequent rotation of Canadian troops brought Roman Catholic Padres Jean-Pierre Guay (priest) and Yvon Pichette (pastoral associate) to Visoko. Together, they profoundly demonstrated the strategic value of continuity in operational ministry. These chaplains further facilitated the networking that had begun during the previous rotation among the religious leadership of these ethno-religious communities still at war. The ministry of hospitality continued in sincerity, as did the discourse with the residual effect of establishing still greater trust. In Split, Croatia, I had but a brief conversation with Padre Pichette as segments of our two contingents literally crossed paths in rotation. I informed him of the interest in intercommunal dialogue demonstrated by Asim Azdahic, Senior Imam of the greater Visoko/Kiseljak area. In the coming months, Padres Guay and Pichette expanded the dialogue to include the local Serbian Orthodox priest, Miroslav Drincic, thus making contact with local and regional leadership of the three faith communities of Central Bosnia.

The « Rapport des aumôniers du GT 2 R22eR » (Chaplain Report) submitted to the Commandant du Secteur du Québec de la Force Terrestre

(SQFT) succinctly outlines the goals for their ministry as the seeding of reconciliation among the three faith communities. Their strategy clearly emphasized the need for nurturing trust in the building of relation among the local religious leaders and their respective faith communities. As outlined in their report, they desired to:

- Understand the situation relating to the various communities of faith of persons residing in central Bosnia, and their humanitarian needs;
- Develop partnerships with the spiritual leaders of the communities;
- Attempt to find common ground between the factions by way of small iterative steps; and,
- [Engage in] ecumenical prayers for peace.[11]

In support of this vision, visits were made to the various Franciscan and Muslim leaders of Visoko and Kiseljak[12] and, more significantly, to the Serbian Orthodox priests of Zenica and Kraljeva.[13] As relation moved beyond introductions, Father Guay's empathic listening enabled these leaders to share the pain of their war-ravaged communities and fractured fraternity, stories all strikingly similar.[14] It is through the ministry of hospitality that the sense of "safe space" is created, enabling such sharing to occur, a companion to trust for those who are living in the midst of conflict. Hospitality also shares a sense of gratuity, a kind of graciousness that is not measured in a *quid pro quo* manner. There is an abundance that invites thinking about new possibilities. Emanating from the gratuitous nature of hospitality is the caution not to make it into something other than what it is. "Hospitality has its own place and its own priority. Part of the experience of its gratuity, then, is that it is not used for any extraneous purposes: it is there for its own sake."[15] Trust grows out of such hospitality between the operational chaplain and the religious *other*.

Such private sharing and public participation in *les rencontres de prière*[16] (prayer meetings) with all three faith group leaders and their communities inspired Padres Guay and Pichette to propose a « Célébration œcuménique pour la paix »[17] (interfaith celebration for peace) involving these leaders and members of their respective faith communities. Upon learning of the precedent-setting nature of this multi-faith celebration in a virtual theatre of war, Radio Canada producers of the Québec-based programme, « Jour du Seigneur » inquired and received permission to videotape the scheduled 29 June 1993 event, which aired in Canada on 12 September 1993.[18]

Erudite individuals, these faith group leaders immediately fathomed the symbolic significance of such an interfaith celebration to the long road of *reconciliation* that lay before them and their communities. At the suggestion of Guay, each cleric agreed to offer a symbol from their faith tradition to present to the assembled, representing the richness of each religious confession. These symbols were to dwell permanently in the tent that served as the chapel for CANBAT II,[19] testifying to the mounting goodwill among these leaders despite the conflict, past and present, which had engulfed their collective communities. Individual relation-building between the chaplains and the religious leaders engendered the beginnings of trust among these clerics, trust enough to join together in a symbolic event for the greater good of all of their people, regardless of faith tradition or orientation of their respective political leaders. To provide a greater sense of how relation was created in this particular instance, the following extended excerpt from the celebration liturgy describing the presented symbols is offered:

> Spiritual leaders from central Bosnia, coming from various religions represented here today, will exchange treasured symbols, each representing the confession of their religious faith.
>
> First, the Imam Asim Azdahic, Muslim priest, President of the Islamic Community of the region, will give Padre Guay the Islamic symbol, which will always reside in the tent of the chapel in our community of Camp Visoko.
>
> Silence and Islamic song.
>
> The Orthodox priest Miroslav Drincic, in charge of the Zenića parish for the last 22 years, presents an icon of the Virgin Mary, Mother of God. Let us recall the importance of the Gospa throughout the world, and particularly in the last twelve years at Medjugorje.
>
> Orthodox song.
>
> Tomorrow, 4 October, celebrates St. Francis of Assisi in Italy, and the Franciscans have grown the Christian communities all over the former Yugoslavia. Srebrenica erected its first chapel 700 years ago.

> Twice already, Assisi was the meeting location for the spiritual leadership of the world where they prayed for peace. Father Paolo de Visoko presents us the effigy of St. Francis. This will serve as an important symbol among us as a model artisan of peace.
>
> Prayer of St. Francis from the Blue Book.
>
> Padres Guay and Pichette present a crucifix as a symbol of our communion recognizing the communities of believers that reside in Bosnia. It is also a symbol of our journey among them.[20]
>
> Crucifix.
>
> Comments by Major Guay.

Also, poignantly symbolic of the hope that remained for war-torn Bosnia was the intentional involvement of children from the participating faith communities, life-giving signs of the future. The celebration continued:

> Children place baskets on the table.
>
> The children bring flowers.
>
> Songs.
>
> Comments: As long as there are children, there is hope, regardless of the suffering and the obstacles ... as long as there are roses and lilies in Bosnia, there will be a sign of life that will always be growing. Let us see and hear the children sign about life and peace.[21]

Such occasions give amplification to both the possibility and efficacy of speaking with one religious voice "so that others may hear a 'religious voice,' rather than a partisan voice, especially in an area where the ethnic and religious fault lines are so closely aligned,"[22] aiding the resolution of conflict via the emergence of religious pluralism.

Notably, it was the embracing of relation exemplified by Padres Guay and Pichette with all three faith group leaders and their faith communities that precipitated the establishment of dialogue between the Serbian Orthodox priest and his counterparts. Based on an engendered trust, the chaplains' openness to relation with these religious leaders and their communities created an environment where inclusivity of the estranged *other* was approximated. The modelling of hospitality and acceptance of the religious *other* by these padres precipitated similar desires within the other religious leaders, initiating encounters among them that had not existed since a time prior to the commencement of hostilities. Modelling of this nature is understood to be *mimetic*[23] in the sense that witnessing such openness to relation in another – the genuine desire for their well-being, the willingness to recognize the humanity of the *other* – plants the seed of desire for the same within an individual or group. This atmosphere of trust emanating from the building of relation cultivates an environment conducive to the emergence of another perspective. Rising out of the chaos of conflict comes a nascent desire for renewal. Such seeding of reconciliation scratches the surface of belief systems and ideology, the wellspring of conflict.

In regards to this case study, the nurturing of trust made way for renewed mutuality, fostering the beginnings of a reframing of relation between the Serbian Orthodox priest and his Croat and Muslim counterparts. The mutuality of the *self* and the *other* brought into view greater receptivity, moving the Serbian Orthodox priest, and by extension his community, away from their isolation toward inclusivity with the *other*. The interfaith celebration offered an opportunity to underscore the reality that security was to be found in relation. The beginnings of trust made possible by encounters of this nature aids in the re-humanizing of the *other*, when other forces purposely manipulate religious belief and maliciously manufacture distorted historical accounts as a means to demonize and dehumanize the *other*. These religious leaders came to trust Padres Guay and Pichette as individuals who manifested genuine concern for their well-being and that of their people. Such trust opened the door to their willingness to share in an interfaith celebration for peace when their communities were still engulfed in overt conflict.

The richness of symbolism[24] embedded in this ritual celebration and its effects merits a degree of articulation. As a formal and constructive ritual, the inclusiveness of the interfaith service temporarily suspended the social structures and roles that had perpetuated alienation and conflict in greater Visoko. As a ritual of inversion, the grip of local power structures on social relations among these communities was muted, thus tipping the power structures on their head, if only for the duration of the ritual event. Mimetically modelled by Guay and Pichette was a passageway to a new reality of social structures and relation. Of particular note was the presentation of icons from each faith community: « un symbole Islamique; un icône représentant la Vierge, Mère de Dieu (orthodoxe); et l'effigie de St. François (Franciscains) »[25] left to reside together in the camp chapel, a glimpse into the liminal space of a transforming relation and budding unity. Transcending the inherent symbolism of the individual icons was the intentionality of the three religious leaders to, at least for the moment, redefine their worldviews and identities. Visible also was the message of connectedness and re-humanizing of the *other*, which for some may have registered only subliminally. A dimension of symbol and ritual in peace-building is the depth of meaning imparted by an event that may initially evade the participant's ability to articulate. Nonetheless, enshrouded in symbol, the seed of truth at some level takes root, slowly germinating over time.

Reinforcing their shared humanity was Padre Guay's utilization of an expansive hand-painted mural that served as a backdrop to the makeshift chancel area: the symbol of a mature tree with large branches, full foliage and exposed roots. He commented:

> This tree represents all the roots of the same God, from which a unique life has given the world each and everyone one of you. Regardless if we are Muslim, Jewish, Orthodox, Christian or of all other protestant or catholic religions, we form the trunk together. The trunk is represented here today by the prayer of this unified family who pray to the same God so that each of our gifts, each of the religions, each of our lives can grow like these leaves in the sun which shine of the same life, and of the same faith.

Guay's emphasis on the common heritage of the Abrahamic faith traditions was yet another unifying message of kinship and rootedness, reinforcing the humanity of the *other*, pluralism and unity, something that had been distorted by the war. He reminded his listeners of the identity of the *other* aside from that of the conflict. The shared celebration contravened the tendency of the conflict to rigidly centre on one aspect of the *other's* identity.[26] Regardless of religious orientation, religious/cultural ritual and symbol present to operational chaplains an occasion to connect with local religious leaders in a manner not afforded to other contingent members. In theatres of operation, such time-specific events, although short in duration, occur often. The vision of transcendence is for renewed relation in ritual to "carry over into non-ritual time and space."[27] Such events cannot come to fruition without the engendering of trust.

Issuing from such transcendence in the renewal of relation, these three clerics recognized the necessity of speaking with a unified religious voice in contrast to the ubiquitous partisan one that had torn their communities apart. Enough trust developed for members of each faith community to participate in the interfaith celebration. Accentuating the vital importance of open ethno-kin connections for future generations was the intentional presence of their youth.

This chapter has been offered as a means of examining the role of trust in relation building among chaplains and their religious counterparts in theatres of operation. No other contingent member brings such expertise to the field of conflict. Given the climate of today's conflict and its ever-increasing aberrant religious dimension, it behooves commanders to reflect on the strategic merit of engendering trust with local religious leaders as one aspect among many in resolving longstanding and ongoing conflict.

ENDNOTES

1 Miroslav Volf, *Exclusion and Embrace: A Theological Exploration of Identity, Otherness, and Reconciliation* (Nashville, TN: Abingdon Press, 1996), 199. Volf, a native of the former Yugoslavia (Croatia), is now an American citizen teaching at Yale Divinity School.

2 R. Scott Appleby, *The Ambivalence of the Sacred: Religion, Violence and Reconciliation* (Lanham, MD: Rowman & Littlefield Publishers, 2000), 61.

3 Liora Danan and Alice Hunt, *Mixed Blessings: U.S. Government Engagement with Religion in Conflict-Prone Settings*, A Report of the Post-Conflict Reconstruction Project (Washington, D.C.: CSIS Press, 2007), 1.

4 Ibid., 52.

5 Ibid., 26.

6 Chaplain (Colonel) William Sean Lee, ARNG, Lieutenant-Colonel Christopher J. Burke, USAF, and Lieutenant-Colonel Zonna M. Crayne, ANG, "Military Chaplains as Peace Builders: Embracing Indigenous Religions in Stability Operations," U.S. Army Peacekeeping and Stability Operations Institute (2004), as cited in Danan and Hunt, *Mixed Blessings*, 27.

7 John Paul Lederach, *Building Peace: Sustainable Reconciliation in Divided Societies* (Washington, D.C.: United States Institute of Peace, 1997), 41.

8 For three months, CANBAT II provided escort protection for humanitarian aid convoys destined for "UN Declared Safe Havens" in southern Bosnia and also similar protection for contracted civilian workers endeavouring to revive the infrastructure of Sarajevo, namely electricity, water and sewage.

9 Captain S.K. Moore, "Personal Journal of OP CAVALIER (UNPROFOR)," 9 November 1992 - 15 May 1993, 116-118.

10 Miroslav Volf contends that parties in conflict often call for *strict justice*, i.e., the balancing of accounts of past injustice as a precursor to reconciliation. Volf is persuaded that such rigidity precludes any movement between victim and perpetrator. His understanding of human interaction suggests that there is too much injustice in an uncompromising struggle for justice. The *will to embrace* justice is transcended in the sense that willingness within the *self* emerges for initial engagement with the *other*, recognition of the humanity of the *other*. The *will to embrace* justice is affirmed. Inherent in its promise is an eventual attending to injustice as former enemies move from entrenched positions to taking small steps toward renewed relation. Volf speaks of the willingness of usually the wounded party to see the good in a gesture of the *other*, often against a backdrop of the tremendous hurt and suffering known to conflict. Movement toward the *other* often precipitates a reciprocal reaction. The *will to embrace* involves not only seeing the *other* from a different perspective, but also seeing the *self* from the perspective of the *other*, something Volf calls "double vision."

11 Major J.-P. Guay et Capitaine Y. Pichette, « Rapport des aumôniers du GT 2 R22^{e}R », Annexe B, A 1630-1 (Aumônier), octobre 1993, B-1.

12 Ibid.

13 Ibid.

14 Telephone conversation with Major Jean-Pierre Guay, Aumônier-adjoint, Quartier Général SQFT, Montréal, Québec, 16 September 2005.

15 Robert Schreiter, *The Ministry of Reconciliation: Spirituality and Strategies* (Maryknoll, NY: Orbis Books, 2002), 90.

16 Guay et Pichette, « Rapport des aumôniers », B-2.

17 Major J.-P. Guay, Projet production pour l'émission « Jour du Seigneur » de Radio Canada, Célébration œcuménique pour la paix, « Note de service », 1110-1 (Aumônier), 26 juin 1997, 1.

18 Guay et Pichette, « Rapport des aumôniers », B-4.

19 Guay, Projet production, 9. In conversation with Padre Guay (16 September 2005), the unprecedented nature of the interfaith service was amplified. In concurrence with the much-needed expression of fraternity and solidarity among the religious community

leaders of multiethnic greater Visoko, Miroslav Drincic (Serbian Orthodox priest) agreed to participate. As is the policy of chaplains, permission for the interfaith service from Lieutenant-Colonel Desjardins (CO, CANBAT II) was sought and granted. The location of Zenica was behind Serb lines, which posed a degree of danger for this priest and community leader. Three Cougars (military vehicles) were sent to safely transport Father Drincic and some children from his parish to the Canadian compound for the multi-faith service. Padre Guay also related that Imam Azdahic was inexplicably late for the interfaith celebration. He arrived just prior to the benediction and was immediately invited forward to read the earlier agreed-upon text from the Koran in Arabic. It was learned following the service that certain members of the imam's community were opposed to his participation, causing him to be late.

20 Ibid.

21 Ibid., 9-10.

22 Rabbi Arnie Resnicoff, former Command Chaplain, USEUCOM, US Navy, presently Special Assistant to SECAF and CSAF for Values and Vision, Pentagon, Washington, D.C., e-mail, 6 October 2005.

23 Vern Redekop aptly describes conflict's insatiable appetite for violence by employing the concept of *mimesis*. In its original Greek, *mimesis* carries with it the sense of the *desire* to "mime" and "imitate." *Mimesis* is epitomized in conflict zones through the "mirroring" effect of violence: a victim of aggression is "avenged" by offering violence to the perpetrator, who, in so doing, becomes the new perpetrator. Infused with hatred and venom, such violent reciprocity perpetuates the spiralling effect of the human urge to "strike back twice as hard," hence the oft-quoted idiom, "Violence begets violence…with interest." Major Steve Moore, "Military Chaplains Seek to Bridge Conflict," *Emmaus: A Newsletter For Friends of the Canadian Council of Churches* 3, no. 2 (Winter 2006-2007), 5. In offering balance to mimetic structures of violence, Redekop presents *mimetic* structures of blessing: empowering, leading to creative and ever-expanding options, and oriented toward life. Rather than mirroring violence, these structures echo a reciprocal action of intending, wishing for, and acting toward the mutual well-being of the *self* and the *other*. For further background on *mimetic* structures of violence and of blessing, see Vern Neufeld Redekop, *From Violence to Blessing: How an Understanding of Deep-Rooted Conflict Can Open Paths to Reconciliation* (Ottawa: Novalis, 2002), 61-84, 161-172, 255-284.

24 The theory presented here with respect to symbol and ritual originates with Lisa Schirch, *Ritual and Symbol in Peacebuilding* (Bloomfield, CT: Kumarian Press, 2005).

25 Guay, Projet production, 9.

26 Schirch, *Ritual and Symbol in Peacebuilding*, 154.

27 Ibid.

CHAPTER 7

TRUST: THE CRUCIBLE OF SUCCESS IN COMMAND

Colonel Bernd Horn

The often-used phrase "just trust me" is rather ironic and paradoxical. Its use is meant to provide confidence and reassurance, yet, if it has to be used at all, it is clear that a trust relationship does not actually exist. Normally, it is indicative of a situation in which individuals have no choice but to put their fate in the hands of someone else. If trust were present, there would be no need to speak of it. Broadly defined, trust is a bond of faith between two people or a number of people that has been built through shared hardship and experience, or over a period of time, in which a mutual understanding and confidence exists.

Trust is particularly important in the military. It knows no rank or appointment and no words or discourse can create it – it is a bond formed from action and actual deeds. Trust is dependent on building credibility through what one does, not what one says. It is born from the acid test of real operations and activities where military personnel, regardless of unit, formation, branch or environment/service rely on one another for safety, if not for survival. Whether standing sentry, preparing munitions or mission essential equipment for operations, conducting operations, or supporting them, trust is important in developing the confidence to focus on the mission at hand without worrying unduly about the peripheral details.

Paratroopers are a good example. They rely on the air force, particularly the pilots, navigators and flight crew, to get them safely to their destination. More important, they rely on the parachute packers and riggers to prepare their equipment safely and competently, as well as the jumpmasters to ensure that they are properly dressed and dispatched from the aircraft. Finally, they rely on one another to conduct safe drills in the air and to provide the necessary teamwork and skills when on the ground. The bonds of trust that are forged over time are not built on rank or affiliation, but rather are developed out of shared experience. It should come as no surprise that the closer the working relationship, proximity of support, or level of shared hardship and experience, the greater the bonds of trust that develop.

As important as trust is in all military ventures, nowhere is trust more essential than in command. Trust is the crucible of success. Without it, there can be no initiative or timely decisions and the military's most valuable and effective resource, namely its people, can easily become marginalized.

Command, in accordance with the Canadian Forces' (CF) accepted NATO definition, "is the authority vested in an individual of the armed forces for the direction, co-ordination, and control of military forces."[1] The essence of command is the expression of human will, an idea that is captured in the concept of commander's intent as part of the philosophy of mission command. Mission command is a command philosophy that promotes decentralized and timely decision-making, freedom, speed of action and initiative that is responsive to superior direction. It entails three enduring tenets: the importance of understanding a superior commander's intent; a clear responsibility to fulfil that intent; and, timely decision-making. At its core, the fundamentals of mission command include: unity of effort; decentralized authority; trust; mutual understanding; and again, timely and effective decision-making.[2]

Key to mission command is the concept of the commander's intent. It is the commander's personal expression of why an operation is being conducted and what he or she hopes to achieve. It is a clear and concise statement of the desired end-state and acceptable risk. Its strength is the fact that it allows subordinates to exercise initiative in the absence of orders, or when unexpected opportunities arise, or when the original concept of operations no longer applies.

As such, for mission command to be successful, the command climate must encourage subordinate commanders at all levels to think independently, take the initiative and not be risk averse. The strength of mission command, after all, is the ability of commanders at all levels to react quickly to developing situations in an ambiguous, complex, fluid and chaotic battlespace. In such an environment, it is critical to gain and maintain the initiative. Delay in decision-making can have serious consequences. As a result, subordinate commanders, acting within the commander's intent, must make decisions and take action. This entails trust. The superior commander must trust his or her subordinates to act, to act in accordance with the commander's intent, and to make reasonable decisions regardless of the circumstances that the subordinate commander finds himself or herself in (which in an ambiguous and chaotic security environment may not necessarily be those the superior commander originally envisaged). This level of trust is critical for mission command to function properly. The superior must minimize control mechanisms and allow his or her subordinates the necessary freedom of action and initiative to achieve the necessary effect.

Similarly, the subordinate must have confidence that he or she has not been given an unachievable task and must trust his or her superior to provide the necessary direction, guidance and resources to successfully achieve the assigned mission. Importantly, for the subordinate to fully exercise initiative and accept the necessary degree of risk, he or she must be able to trust the superior to provide the necessary support should errors occur (other than those arising out of malicious intent or negligence).

The issue of trust in command was never as stark as during the 1990s, dubbed by many, but most recently by General Rick Hillier, the Chief of the Defence Staff, as the "decade of darkness." During this period, a complete absence of trust existed and it marked the 1990s as probably the darkest period in CF history. Overcome by concurrent and dramatic social, political, international and technological changes, the CF officer corps, due to its longstanding anti-intellectualism and reliance on an experiential approach, imploded. The societal revolution that saw Canadians transition from an attitude of deference to authority to defiance of authority compelled the military, which for so long during the Cold War was left relatively alone to man the fortress of democracy, to account for its conduct and spending practices.[3] Concomitantly, the fall of the Berlin Wall marked the end of the Cold War and much of the rationale for the CF's existence, the foundation of its entire structure and methodology, became defunct. New operations in a turbulent and unfamiliar world that existed with enemies that could no longer be templated, and a voracious press that had instantaneous access to news feeds and found their way onto every battlespace, soon overloaded the capabilities of a Cold War-centric officer corps. The torture killing of a Somali detainee, an attempted cover-up at the highest level, a barrage of allegations of systemic harassment in the CF, disregard by the chain of command for the rank and file, political manoeuvring, a widely perceived elite parachute regiment that was ostensibly out of control – just to name a few of the stressors crushing the CF officer corps – created an environment of risk aversion and distancing, or in simple terms, a "cover your ass" approach by leadership, particularly at the most senior levels.

Subordinates at all rank levels found themselves abandoned by their superiors. Mere allegations ruined careers, for in this climate, no senior leader would dare stand in the way of an accusation, much less defend an individual, lest he or she be dragged into the fray as an "apologist" or be seen as attempting to defend an action that was alleged to have occurred. On operations, vague direction and a reluctance to provide timely decisions often left field commanders, whether officers or senior non-commissioned members (NCMs), with the need to make the necessary decisions on their own with the full knowledge that if something went wrong, they alone would be responsible as it had become clear that higher commanders would not protect subordinates.[4] Not surprisingly, trust plummeted and surveys repeatedly revealed that neither CF personnel, nor the public at large, had faith in the senior leadership of the CF.[5]

One example still stands out. In the early 1990s, in Bosnia-Herzegovina (BH), the rules of engagement (ROEs) were not clear with respect to the actions that could be taken to defend civilians who were being victimized by host-nation police, military, para-military, state sponsored warlords or militia groups. Pleas for clarity to the chain of command were never answered. Yet, at staff college, a Canadian "two-star" general, who was a sector commander in BH, as part of his address to the class, berated the majors and lieutenant-colonels for the reticence of their peers to "take action" in theatre. He criticized the fact that they always wanted explicit orders and directions. This statement alone was indicative of the level of trust that existed. Nonetheless, the general was quickly inundated with examples of why this was so, all stemming from the previous discussion concerning the lack of trust.

Amusingly, he was asked his opinion on the actions that should be taken if troops witnessed a civilian being assaulted or attacked by forces as described above. He quickly and confidently asserted that clear action should be taken by the troops on the ground. He was then reminded that the ROEs did not cover that response. Not losing any of his confidence, he roared that there was always the moral imperative. The questioner now drove home the point that many were already aware of: he asked the general how it was, if it was so clear, that when he was commander, he and his headquarters refused to issue any clarifying orders or direction and consistently left those asking for authority to act hanging in the wind. Under his tenure, his subordinate commanders were continually forced to act on their own without any protection or, alternatively, to decide not to act at all, moral imperative or not. The confident general visibly lost some of his composure and blustered on, clearly shaken by the unrelenting attack by his juniors.

The point of all of this is that the atmosphere of careerism, "ass covering" and abandonment, if not the complete abrogation of the responsibility to look after and support subordinates in the 1990s, eroded trust to the point that it impacted CF effectiveness.[6] Individuals became reluctant to do their jobs, much less take any form of risk. Supervisors refused to objectively rate their subordinates or counsel individuals fearing that if they did so they would become the target of "harassment" complaints and then would subsequently be "cut loose to deal with it themselves." One general admitted that an "allegation was all that it took to ruin a career." Ironically, although this senior infantry officer openly admitted that he was conversant with the problem, he did nothing. In fact, he is remembered by many as one of the most careerist and deceitful moral cowards of the period (he is in good company however).

Nonetheless, the renewed effort at professional development and a concerted attempt to rebuild the officer and senior NCM corps (i.e., Officership 2020 and NCM Corps 2020), as well as seminal doctrinal publications such as *Duty with Honour: The Profession of Arms in Canada*, have made a tremendous difference. Building trust in a command, however, takes both time and effort. As stated earlier, it is a function of actions, consistent actions, which build the necessary levels of trust that are so crucial to ensuring an effective command team and command, regardless of the level.

For example, when I took command of my infantry battalion, I made a point of articulating my command mentality. I emphasized that I believed that we should use the peacetime environment to learn and instil the proper command attitude, namely mission command and orders. I explained that this meant that activity would be conducted through decentralized execution based on mission orders, a technique that allows subordinates maximum discretion and freedom of planning and executing tasks within the scope of the commander's intent. Quite simply, superiors would provide subordinates with the necessary direction, guidance, information and support – the "what" – and subordinates would have the freedom of action through disciplined, reasoned initiative to provide the "how."

I stressed the necessity of action (within the context of the commander's intent) in the absence of orders. I believed timely initiative was of far greater value than the risk of potential problems made by bad decisions. I also added that careful planning and preparation were always essential, but once committed to a plan of action, decisive execution (i.e., speed and violence) was critical. The theory sounded great, but I realized that implementing it would require substantial effort. As discussed earlier, such an approach implies risk, something subordinates, particularly those coming out of the "dark decade" of the 1990s, were hesitant to accept or, more accurately, were willing to believe their superiors would accept.

In essence, all of those serving in the land forces at this time were conscious of the army's decision to adopt mission command. Most of us, however, were also aware that theory and reality had not yet coalesced into a single entity – the issues of trust and risk acceptance (so dangerous to careers) had not yet been aligned. It was still very much personality-centric. Now being in command, however, I decided that I would make an honest effort and actually "walk the talk." Within the unit, I stressed that we would accept honest mistakes as a method of learning. As long as there was no malice, negligence or apathy involved, a mistake would be treated as an opportunity to learn. I explained that rather than waste time on assigning blame, we would spend our energy in determining how or why a mistake occurred and what actions we would have to take to ensure that the likelihood of recurrence was minimal.

In addition, I realized that everyone had their own way of completing tasks and sometimes the hardest thing to do is to allow someone the freedom to do it their own way. Nevertheless, I undertook, as the commanding officer (CO), to provide the necessary direction and guidance (i.e., my intent) and I always made myself available if subordinates required additional assistance. Critical to this was maintaining an approachable demeanor. As such, I announced, reiterated and practiced an open-door policy. I also performed daily walk-abouts, visiting and talking to everybody, and at orders groups or any other venue, repeatedly reminded everyone of the importance of a clear and timely passage of information through clear communication channels. Having done that, as hard as it was, I let them go. Part of the difficulty is that in the army, leaders are imbued with a cultural mentality that advocates "lead from the front, get it done, mission success at all costs." Moreover, leaders, particularly in the combat arms, are usually "Type A," aggressive and very self-confident, normally empowered with the realization that they, and often they alone, hold the secret of how to best get something done. Therefore, the easiest thing in the world is often to give the task, as well as the detailed explanation of how to do it. Such an approach also provides a perceived assurance of task completion, very important to those who are risk averse. In practice, however, this style is flawed for it dampens initiative and fails to utilize the imagination of those around you. In the end, more often than not, subordinates will amaze you with their innovative solutions and plans. But in this case, it did not happen, at least not right away.

Amazingly, at first, I found that individuals were reluctant to use their initiative, wanting to get authority and clearance on the most minor and trivial details and decisions. They were obviously risk averse and not accustomed to having the latitude to work within the full capacity of their positions. It took constant reinforcement, that they were empowered to make the necessary decisions within their parameters, to make them more at ease. Most important of all, it took supporting their decisions and taking responsibility for any miscues or mistakes that transpired (which were actually very few).

Initially, it was actually all quite comical. When a perceived mistake was made, when something was "dropped" (i.e., a resource request or report was submitted late or incomplete) or when higher headquarters queried something, the immediate action of the operations and training staff and/or the administrative support staff was to appear at my door to explain how they were not responsible for the error and to offer their view on who was. It only occurred two or three times. I cut them off each time. I stated I was not interested in allocating blame, rather, I wanted to know how the miscue happened and what we were doing to ensure it would not reoccur.

This approach solved the problem. Overall, I tried to ensure my actions reinforced the notion that I was willing to accept risk and that mistakes were not career-ending catastrophes. After these initial experiences, my staff brought problems to me in a calm rather than excited state and explained what had happened, why, and what had already been put into place (through their immediate application of initiative) to ensure that it would not happen again. Moreover, the sense of empowerment permeated all that they did and they became the most credible and effective operations staff in the formation. The brigade commander later conceded that his staff dropped an inordinate amount of complex "fastball" tasks on our battalion because he knew the unit would deliver in the crunch.

But this approach actually achieved much more. By not entertaining or engaging in the "blame game," unhealthy competition between sections and/or individuals was curtailed. Individuals could not, and soon afterwards would not, try to capitalize on or gain from the mistakes or miscues of others. In fact, a very cooperative environment developed where teamwork flourished. In addition, the emphasis on actually learning from mistakes, rather than punishing them, created a learning environment where all could profit (from an individual professional development perspective, as well as organizationally).

Such was only possible because of the professionalism of the staff and unit personnel. But more than that, it was a result of trust, namely the trust that I had in my subordinates to do the right thing and make the necessary, timely decisions in accordance with my intent. It was also a product of the trust that my subordinates had in me, namely that I would not give them a task that they could not possibly accomplish, that I would provide them with the necessary resources to do it, and that I would support and protect them should mistakes occur. Trust became a vital component in the success of my command.

In the end, trust is critical to CF effectiveness. Without it, commanders cannot possibly leverage their most valuable resource, the men and women of our great institution. For those who endured the dark days of the 1990s, the importance of trust in command needs no explanation – they witnessed the utter chaos, loss of morale and subsequent ineffectiveness that its absence created. Conversely, to realize that when trust flourishes it truly is a combat multiplier, one need only talk to troops in the CF today, particularly on operations, and assess the impact that our relatively small force is having internationally.

ENDNOTES

1 Canada, *Command* (Ottawa: DND, 1997), 4.

2 See Bernd Horn, "Institutional Leadership: Understanding the Command, Leadership, Management Nexus," in R.W. Walker, ed., *Institutional Leadership in the Canadian Forces: Contemporary Issues* (Kingston: CDA Press, 2007), 99-107, for a more detailed discussion of command.

3 See Peter C. Newman, *Canadian Revolution 1985-1995: From Deference to Defiance* (Toronto: Viking Press, 1995). For an account of the impact of the changes to the CF in the 1990s, see Bernd Horn and Bill Bentley, "The Road to Transformation: Ascending from the Decade of Darkness," in R.W. Walker, ed., *Institutional Leadership in the Canadian Forces: Contemporary Issues* (Kingston: CDA Press, 2007), 1-25.

4 As incredible as this may seem, one need only speak to those who endured the 1990s to appreciate the level to which trust plummeted. See alternatively, Canada, *Debrief the Leaders* (Ottawa: DND, 2001).

5 Bernd Horn, "Wrestling with an Enigma: Executive Leadership," in Bernd Horn, ed., *Contemporary Issues in Officership: A Canadian Perspective* (Toronto: CISS, 2000), 123-144. This reference includes statistics and references to polls taken at the time that demonstrate the depths to which trust plummeted during this period.

6 This reality led to the Officership 2020 project under Lieutenant-General Roméo Dallaire, which was ultimately responsible for creating a renaissance in officership and professionalism in the CF. Its recommendations are today paying great dividends. Stemming from this project were such concepts as the CF University (realized in the Canadian Defence Academy (CDA)), a leadership institute (realized in the Canadian Forces Leadership Institute (CFLI) under CDA), and leadership and profession of arms doctrine (realized in the four doctrinal leadership manuals, an ethics manual, and the seminal *Duty with Honour: The Profession of Arms in Canada).*

CHAPTER 8

CONFRONTING REALITY: A CASE STUDY OF BUILDING TRUST AT THE COMPANY LEVEL

Major Tony Balasevicius

Leadership means different things to different people, and from an historical perspective, the concept has been defined in many different ways.[1] Most definitions derive from personal experience and these individual perspectives may or may not be linked to the direct practice of leadership. In fact, according to scholar Gary Yukl, "Leadership has been defined in terms of traits, behaviour, influence, interaction patterns, role relationships, and occupation of an administrative position."[2] He further states, however, "that most definitions of leadership reflect, to some extent, the assumption that it involves a process whereby one person exerts intentional influence over other people to guide, structure and facilitate activities and relationships in a group or organization."[3]

From a military perspective, this definition appears to have merit, as the concept of effective military leadership supports the position of influence. For example, the recently released *Leadership in the Canadian Forces: Conceptual Foundations* defines leadership as "directly or indirectly influencing others, by means of formal authority or personal attributes, to act in accordance with one's intent or a shared purpose."[4] In order to influence behaviour in a desired manner, the leader must first establish some type of relationship with his or her subordinates. The strength of this relationship is based on a number of factors, but its foundation is built on the concept of trust.

In its most basic terms, trust can be defined as confident reliance. According to the Center for Engineering and Science, "we may have confidence in events, people, or circumstances, or at least in our beliefs and predictions about them, but if we do not in some way rely on them, our confidence alone does not amount to trust."[5] In this respect, it can be argued that "reliance is a source of risk, and risk differentiates trusting in something from merely being confident about it. When one is in full control of an outcome or otherwise immune from disappointment, trust is not necessary."[6]

The concept of reliance is critical to modern military forces because it relies heavily on a team-based approach to solving problems and completing tasks. This philosophy stresses the idea of reliance within the context of the team, and it is this dependence that brings the need for trust to the forefront of the leader-subordinate relationship. Without argument, developing trust within the military can be difficult. This is due in part to the fact that the leader must constantly monitor and balance the competing demands of the operational mission, the organization's interest, and the needs and desires of his or her subordinates. These opposing demands frequently pull the leader in different directions making the ability to gain and develop trust a constant struggle.

While maintaining trust is in itself a complicated process and constant struggle, attempting to regain it once it has been lost or damaged is far more daunting. This was the situation that faced the leadership of "Hotel" Company, 2nd Battalion, The Royal Canadian Regiment (2 RCR) in the fall and winter of 1993. This chapter will therefore examine the issues surrounding the loss of confidence within the company, as well as the steps taken to rectify the problem. It will then outline the actions taken by the company's leadership to regain the trust of the soldiers and highlight some of the more pertinent lessons that were learned from this experience.

BACKGROUND

In the summer of 1993, 2 RCR had just returned to garrison routine after an operational tour in Bosnia-Herzegovina. While the mission had been long and arduous, the battalion was able to claim a number of important accomplishments. Its successes included a deployment into the Balkans on short notice and in near record time,[7] the establishment of a solid area of responsibility just outside the city of Sarajevo, and the creation of an escort system for many of the relief convoys entering the city. More important, the unit had assisted the United Nations (UN) mission in Macedonia and created the UN enclave of Srebrenica. Although the credit for these accomplishments was given to the unit, these achievements can be directly attributed to the outstanding work of the sub-units (companies) and supporting arms that made up the Battle Group.

"Hotel" Company in particular had what appeared to be a very successful tour as it provided the command element, headquarters and much of the force structure that was instrumental in establishing the first substantial UN presence in Macedonia. This high-profile achievement was all the more gratifying given the complex nature of the mission and the fact that its successful outcome had a significant impact on the operational progress of UN activities in the Balkans. In fact, by all accounts, the general perception of "Hotel" Company was that it was one of the best sub-units in 2 RCR. It had an excellent reputation as a solid organization that had very few disciplinary problems and rarely made mistakes. Consequently, I was extremely gratified when I was told I would be taking over this company after the battalion's return to Canada.

On taking command, my impressions appeared to validate my initial perceptions. The second-in-command (2IC) was new to the job, but was a seasoned officer with good command presence, lots of initiative and solid staff skills. The company sergeant-major (CSM) was fit, intelligent and well liked by the soldiers. The three platoon commanders were all smart, keen and willing to learn. More important, the non-commissioned members (NCMs) were experienced, fit and very hardworking. With a solid team in place, it did not take long for everyone to settle into a routine, and by all appearances, the company was adjusting quite well to the post-deployment circumstances and fresh command personalities. Things are not as they always appear unfortunately, and shortly after taking over command, my perceptions began to change appreciably.

IDENTIFYING THE PROBLEM

Like many significant events, the catalyst to this problem was innocent enough. Lieutenant Walden, a fit, intelligent and aggressive officer who was commanding one of the rifle platoons, came into the office one morning to discuss placing one of his soldiers on counselling and probation (C&P). Before starting the meeting, I asked him to find and bring the CSM with him, as I wanted a different perspective on the matter. Walden had briefed me on the file a few days before and although I was in general agreement with the course of action that he was recommending, I was concerned that C&P appeared to be a drastic step for this particular infraction. As the three of us sat down for the meeting, I wondered out loud whether it would be more appropriate for the soldier in question to be given a lesser form of censure.

To my immediate surprise, Walden appeared somewhat frustrated by the comment and quickly explained that the soldier was already on recorded warning (RW) for a similar infraction, so C&P was the next logical step. I indicated that this reasoning was correct if the soldier was in fact on RW, but I had not seen any documentation to that effect in the file. The CSM interrupted the exchange by stating that he was in agreement with Lieutenant Walden's recommendation and that he was sure the soldier was on RW as he had been in the former officer commanding's (OC) office when it had been administered. I declared that if such was the case, then C&P was certainly warranted and that I was prepared to take the file to the commanding officer (CO). I asked Lieutenant Walden to place the RW sheet in the file and return the folder to me that afternoon. I also asked him to have the 2IC arrange a meeting with the CO through the adjutant.[8]

That afternoon, Lieutenant Walden returned to my office somewhat annoyed and perplexed. He declared that the RW sheet was in fact missing and then went on to state that he had also checked the files of other soldiers he knew to be on either RW or C&P, but was unable to find any of those forms either. Although odd, I initially assumed that the missing forms were little more than an administrative oversight, as many of the unit's files were still being updated from the tour and in the process of being moved around after the battalion's reorganization. I asked the 2IC to check with the adjutant and see if the unit had any information regarding the possible whereabouts of the misplaced paperwork and to have the adjutant hold off on scheduling my meeting with the CO.

When the 2IC completed his inquiries the following day, he reported that the adjutant had no record of anyone being on RW or C&P from "Hotel" Company. Furthermore, he indicated that he had taken the liberty of calling National Defence Headquarters (NDHQ) and discovered that their records indicated at least four members of the company were on C&P.[9] Still confused about what was going on, I called the company administration NCM into my office to talk about the missing documents.

Although somewhat reluctant to say so, this NCM did admit that it appeared someone had accessed the company's personnel records and that the files had been systematically stripped of all incriminating administrative and disciplinary paperwork. He went on to say that he had also just informed the chief clerk about the matter and that the unit was now looking into the situation. Shocked by this revelation, I asked the administration NCM why this issue had not been brought up sooner. He emphasized that he had only just become aware of it himself. When pressed about how someone could have gained access to the files and how the records of those files could have been erased from the unit logs, all he would say was that "you are new and do not understand." He then stated that he could not say anything more because he was afraid of what might happen to him.

Although it was evident that something was wrong, I was still trying to figure out what it was when I called Lieutenant Walden and the CSM back into my office. Unfortunately, both were as stunned as I had been, and neither of them had an explanation for the NCM's admission or concerns. In an effort to obtain more information, the CSM suggested that he do a "walk-about," to talk with the troops and see what he could pick up. He came back the next day and acknowledged that there was undoubtedly something wrong but that no one was prepared to speak up. Never having experienced this type of a situation before, I was at a loss as to what I should do next.

I knew that I had to go to the CO about this as soon as possible, yet as it stood, I had very little to tell him other than some paperwork was missing and that the soldiers appeared frightened to talk about the activities of the sub-unit. Moreover, I was not even sure that the two facts were linked. That being said, sorting out the paperwork was now of less concern than the soldiers' silence, as experience told me that if soldiers are unwilling to talk to the chain of command, something "big" was at play. Understanding this, it was clear that getting to the problem meant first getting the soldiers talking again.

In general, soldiers are not shy about speaking up when they have issues with the chain of command unless leadership has shown a pattern of suppressing such initiatives. The other possibility was that the problem was among the soldiers themselves. The worst-case scenario was that it was a combination of the two. Regardless, if soldiers feel that they cannot trust their superiors to help them, they will not come forward. I realized that in order to get at the essence of the problem, I had to create an environment where the soldiers who wanted to speak felt comfortable doing so. More important, I knew that soldiers would not give up information if there was any chance that they would be found out, thus information gathering had to be completed in a manner that guaranteed anonymity.

To this end, I instructed the CSM to schedule interviews with each person in the company. I told him that just before a soldier entered the office, they were to be briefed on the purpose of the interview and that it would last for five minutes. The soldier could say something or nothing, but at the end of five minutes, the soldier would leave.[10] My hope was that going through all of this would demonstrate to the soldiers that I was serious about fixing the problem. Fortunately, a number were confident enough to speak frankly; all they needed was a venue and someone they could trust that was willing to listen to their concerns. To a man, however, they were not prepared to provide formal allegations. As one soldier pointed out, "I am glad that someone is finally looking into this.... I can give you the information but that is as far as I will go."

In putting the various bits of information together, it became clear that there was a group of five soldiers who were systematically terrorizing the company. This group was backed by about five others who appeared to be "sympathizers" and who would associate on occasion with the "problem" soldiers. Statements made by some indicated that the major activities of this group ranged from drinking while on observation post duty, dealing in drugs, taking drugs, and threatening other soldiers with harm if they talked. Those that exposed the issues believed that if important files were missing from the company, it was likely the result of this group intimidating those people that controlled the documents.

Additionally, both soldiers and junior NCMs suggested that it was common practice within the company to deal lightly with or completely overlook major infractions such as negligent discharges of weapons, lateness and drunkenness. Interestingly, one example that surfaced repeatedly during the interviews concerned a leader who had just been promoted for his "excellent performance" during the tour even though he had been found drunk while on duty. Many believed that the company hid the offence by giving him what was referred to as "platoon punishment."[11]

Unfortunately, by the time all the interviews had been completed and I had sufficient information to go to the CO, it was the day of the unit's change of command ceremony. Although it was probably the worst time to bring up a problem such as this, I felt it was important enough to brief the outgoing CO. After a number of probing questions, he stated that the significance of the situation necessitated briefing the incoming CO before the ceremony.

The new CO was a smart, experienced and aggressive leader who would be a positive influence on the unit in the coming years. After receiving the information, he asked me for my recommendations and then inquired as to what help I would need from him. I responded by saying that I was still working on the plan and that I had scheduled meetings with the military police (MP) and base legal staff later that afternoon. I went on to say that I would have a better idea of my options after these meetings had taken place. He acknowledged my comments and ended the meeting with a "we will talk more about this in the next few days."

DEVELOPING THE PLAN

Interestingly, the CO did not hold off and called me into his office the following morning. I had fortunately gathered the additional information and was more or less ready with my assessment and to outline my plan. I briefed him on the fact that discussions with the MPs had established that at least six members of the company were well known to them and the local police for suspected involvement with drugs and that they were now looking into the matter. As the company's soldiers were unwilling to make formal statements, however, both the MPs and legal staff felt little could be done for the moment.

Based on the fact that we had few alternatives available, I told the CO that my first move would be to bring the company together and inform them that we were aware of the issue. I emphasized that this action would send a clear signal to all that the company's leadership was cognizant of the situation and that it would no longer be tolerated. Based on this assessment, I recommended that no soldier be allowed to move from the company until the situation was sufficiently resolved. The CO agreed with this approach and instructed that I keep him informed with regular updates.[12]

Soon afterward, I recognized that this was going to be a difficult exercise. I was still not sure if I understood the full extent of the problem, and with no one willing to formally come forward, I had few real options. In the short term, I realized that I had to isolate and eventually remove the problem soldiers, but I had to do it in a way that would not simply transfer the problem to another unit. More important, while "cleaning house," I needed to minimize any long-term adverse impact on the company. To do this I knew that I had to regain the trust of the soldiers and re-establish the bonds of the leader-subordinate relationship, which is at the heart of the Canadian army's excellence.

Toward the end of the unit's block leave period, I brought the 2IC and CSM into the office to bring them up to date on the situation and to lay out the plan of action the company was going to follow. I stated that I would start the process by talking to the company the following day. The CSM recommended that we hold off for a few more days as some of the company's soldiers were still on leave and that everyone would be back the following Monday. I said "no," indicating that my target audience was now on the ground and that word would get out to the remainder soon enough. I reinforced my conviction by saying that it was my experience that soldiers pay less attention to what leaders say than to what they do, and in this case, there was no question that our actions, not words, would be the key to our success.

I was about to embark on a house cleaning and rebuilding process. Speaking to the company was the opening round. After having put the soldiers at ease, I launched into my talk by stating that an important aspect of leadership was to know as much about your command as possible and that I had spent most of the past few weeks doing just that. I went on to say that what I had discovered was troubling and I laid out specific incidents that had taken place before, during, and after the company's tour to highlight and reinforce this assertion. I stressed the fact that it appeared there were only a few "bad apples" and emphasized that the people responsible for those incidents had been identified and that in time they would be held accountable for their actions. I then informed the company that the MPs were also aware of the situation and were now actively involved in an investigation of their own. I concluded by saying that while we were dealing with this issue, the company's training routine would continue in earnest as we worked towards, once again, becoming the best company in the unit. I asked for questions and then ordered all leaders to remain behind when the troops departed.

In talking to the company's leaders, I told them that we had a responsibility to create and maintain a safe environment for the soldiers and that for some months now that had not been the case. I expressed disappointment that a situation existed where soldiers were afraid to come forward about a problem and stated that it would not happen again under my watch. I stressed the importance of trust in command and said that if soldiers could not trust telling their superiors about problems in the company, it represented a failure of leadership. I went on to say that this was a failure we were now going to correct in quick order! I summed up the talk by emphasizing that maintaining discipline was the foundation of a good military organization, and that if someone needed to be charged, it would happen. I warned them that a change in the way the company was doing things was forthcoming and that they needed to be prepared for that. Henceforth, we would be an open, fair and highly disciplined organization focused on being the best in the unit. I asked if there were any questions and then dismissed them to their duties.

Interestingly, over the next six months, three of the five soldiers who had earlier been identified as being members of this group decided to leave the military as they were nearing the end of their initial engagements, one requested a transfer for personal reasons, and one was posted to a more closely supervised position within the unit. It was clear that once the group and its activities had been exposed, the group lost much of the leverage that they had had with the soldiers and were quickly placed on the defensive. Having nowhere to hide, many initially attempted to leave the unit by requesting postings to the infantry school and local militia units. These requests were denied since evidence was starting to mount that would eventually be used to hold them accountable for at least some of their activities. Despite the decision of keeping the company intact until sufficient evidence was gathered to move forward with disciplinary proceedings, one door remained open, leaving the military, and most of the group took that option. Although a significant step forward had been taken by exposing the group and its activities, it was clear that much work still needed to be done in convincing the soldiers that they could once again trust the chain of command.

REBUILDING TRUST

According to Canadian Forces' leadership doctrine, "Leaders build and maintain trust through their decisions, actions, and interactions. Hence effective CF leaders at all levels must be able to exercise good judgment in decisions that affect others, show trust and confidence in their subordinates, demonstrate concern for the well-being of their subordinates [while] treating subordinates fairly...."[13] Leaders must also focus on the mission and maintain high standards while engaging in "honest and open communications [with soldiers]...."[14] Although not codified by the company at the time, many of these principles were used by the leadership to good effect in winning back the soldiers' trust.

It was decided that the first step to re-establishing the leader-subordinate relationship within the company was to put emphasis on quality training to facilitate the bonding process. The CO had developed a very aggressive and progressive two-year training plan for the unit that initially emphasized individual and section level training. This formed the foundation for collective training that would ultimately take the unit up to battle group level operations by the spring of 1995. Live fire exercises were scheduled at each stage in the process to provide confirmation that the training objectives had been reached. This ambitious timetable worked to the company's advantage, as training started immediately and gave the leaders and soldiers more time together in the field than would normally have been the case for a unit in garrison routine. This situation served to help re-establish the bonds between leaders and soldiers.

In an effort to remain focused on developing the leader-subordinate relationship, the CSM recommended that the company incorporate section and platoon level competitions into the company's program. He also suggested that all competitions have some type of reward for the winner such as a Friday afternoon off. These recommendations were accepted and at least one competition was run each week. They started at the individual level and included basic soldier skills such as the stripping and assembling of weapons, grenade throwing, first aid, judging distance and marksmanship.[15] These individual exercises quickly progressed to team events such as "march and shoot" competitions where sections and platoons walked at increasingly longer distances, ending with an obstacle course and some type of shooting contest. These activities put emphasis on teamwork and had the benefit of allowing the company to develop cohesion within a competitive environment in a progressive manner by starting at the fire team and working up to the company level.

As the training progressed and skill levels increased, moreover, the company's leadership used the opportunity to show the soldiers that they had trust and confidence in them by giving selected soldiers greater authority. Providing soldiers with greater responsibility was done by allowing qualified NCMs to control company road moves back to the base at the end of an exercise or training event. As road moves back to base were always an administrative event, key NCM command positions such as the platoon warrant officer (WO) and CSM were allocated to responsible corporals and privates. In addition, section commander positions were given to soldiers that had displayed solid leadership potential.

The soldier designated as acting CSM for the move was given a warning order that included a departure time, routes and a time to be back on base. He was expected to assess the situation, come up with a plan, give orders, carry out the necessary coordination, and finally command the convoy back to base. To give more soldiers the experience, the company occasionally broke the convoy into packets with different and independent command structures under the scrutiny of the platoon WO. This activity proved to be an excellent method of evaluating leadership potential while giving soldiers a high profile responsibility. More specifically, it gave the soldiers a clear message that the company's chain of command trusted their abilities.

Not all activities to get soldiers more involved in the decision-making process were done in the field. In one instance, the unit had been given a number of small buildings that were to be used as the new company quartermaster's (CQ) stores. The old stores were small and organized for the efficient control and flow of equipment. One of the major complaints from the troops, however, was that there was no place to work with the equipment once they got it and that this was a major problem during periods of inclement weather. With the new building, there was an opportunity to allow the soldiers to have a say in its design. The soldiers were given an initial basic design to work with and were then asked to provide recommendations on how to improve the layout. Two major recommendations included an area to work in and the ability to access some platoon stores without having to go through the CQ staff. Although there was some initial resistance from the CQ, eventually the benefits of having the platoons take over some of the responsibility for some of the equipment while still holding it in a centralized area were evident. It is fair to say that the soldiers' input transformed the company stores into the best-organized stores in the unit. It did much more, however, for it served to help solidify leader-subordinate relations.

A key part of earning trust is demonstrating concern for the well-being of subordinates, and more important, to treat them fairly. To this end, the company's chain of command went to great lengths to ensure that all decisions concerning the soldiers were made without favour or discrimination. This was all the more difficult than it first appeared, especially as it included the sympathizers of the problem soldiers. Although this particular decision was initially unpopular with some, it was important to show that everyone was being treated equitably. In this respect, unless specific evidence was found to show wrongdoing, all soldiers were treated in the same manner. The concept of fair treatment extended to the nomination of soldiers for promotion, special taskings and career courses.

To supplement these changes, other reforms were instituted to increase soldier involvement in the decision-making process. For example, decisions about the company's fund were taken away from the OC, 2IC and CSM and handed over to a new committee that included the 2IC, CQ and a soldier from each platoon (master corporal or below). Representatives from the platoons were expected to brief the remainder on the outcome of each meeting. This simple action allowed soldiers to not only see how much money was available, but it also empowered or gave them some control over how it was to be used.

In addition to giving soldiers more responsibility, emphasis was also placed on looking out for the well-being of subordinates. To this end, special efforts were made to help soldiers with personal issues. The company also put great emphasis on annual leave. Because the milestones for the unit training plan and exercises were well established, the company was able to provide more flexibility in its leave policy. In this respect, soldiers wanting to take leave around important training periods were given favourable consideration.

It must be noted that being "soldier-focused" did not mean that the company eased off on discipline. In fact, it was quite the opposite. Soldiers still found trouble and were subsequently charged. The main difference, however, was that we had implemented the reforms and the policies on discipline were clearly outlined: everyone understood what was expected of them. More important, the rules were enforced within a process that was open, transparent and fair. In this respect, the key to maintaining effective discipline was setting the rules and applying them with consistency.

LESSONS LEARNED

So what lessons can be taken away from the experiences faced by the leadership of "Hotel" Company during the fall and winter of 1993? The first and most important lesson is that leadership must be prepared not only to recognize a problem, but be willing to deal with it quickly and effectively. All too often the easy approach is taken, that is, transferring the problem individuals thereby making them someone else's issue. Although this solution works in the short term, it does not help the unit, the army or the CF.

Second, it is important to be consistent and to do what you say you will do! In the case of "Hotel" Company, I started off the process by bringing the company together and putting all the facts and issues on the table, largely because I was at a loss as to what else to do. I did not realize that this was probably the most important thing that I did during the rebuilding process. By putting the issues up front, everyone understood that the chain of command was "in the know" and that such actions would no longer be tolerated or go unchecked. Moreover, when the soldiers recognized that the leadership was committed to fixing the problem, it was taken as a sign that command was keeping its word. In fact, doing this was a significant step towards regaining the soldiers' trust.

The full support given to the company by the chain of command was instrumental to the success of our efforts. Interestingly, the CO never got directly involved in the process. He left that to the company's leadership. He always seemed to know what was happening and was always available to provide timely advice when it was needed. The company had full confidence that when it needed something from him, they would get it. This fact alone gave the company's leadership sufficient confidence to proceed with the changes.

Finally, despite the fact that we had a number of major internal issues, it was important to keep everyone focused on the mission and to keep moving forward. In the case of "Hotel" Company, the unit's aggressive training plan allowed everyone to shift back into their comfort zones. Rather than becoming an obstacle to fixing the problem, the high pace of activities became a major part of the reconstruction process.

SUMMARY

It took almost a full year to bring the company to where it needed to be in terms of trust and performance. In the end, all of the policies that were employed by the company's leadership had a direct impact on creating the necessary trust and increasing the company's performance overall. It is important to note, however, that these policies only worked because they were implemented within an environment that was open, fair, consistent and equally applicable to all.

Some will argue that many of these lessons are little more than stating the obvious and that good leadership will ultimately prevent such incidents from happening. This is quite true. Such a statement, however, does not take into account frequent turnovers of leadership within the organization, inexperience, and the self-serving need for some in positions of command to look good in front of superiors. For "Hotel" Company, a number of things went right. A routine issue that could have been quickly forgotten was aggressively investigated. Once the problem had been detected, there was a commitment by the chain of command, both inside and outside the company, to fix, not ignore or transfer, the problem. In this case, the plan had to take into account a small group of troublemakers while protecting the rest of the company. To do this, the sub-unit's leadership had to focus on regaining the trust of the soldiers and then to re-establish the bonds of the leader-subordinate relationship.

In the end, the success of the company's leaders in confronting an issue of trust was based on the application of good leadership principles and common sense. And to those who would argue that many of these lessons are little more than stating the obvious, and that good leadership would prevent such incidents from happening again, all I can say is that they are absolutely correct and ask, how was this situation allowed to happen in the first place?

ENDNOTES

1 Gary Yukl, *Leadership in Organizations: Fourth Edition* (New Jersey: Prentice-Hall Inc., 1998), 2-3.

2 Ibid., 2.

3 Ibid., 3.

4 Canada, *Leadership in the Canadian Forces: Conceptual Foundations* (Ottawa: DND, 2006), online version, http://www.cda.forces.gc.ca/CFLI/engraph/leadership/conceptual/toc_e.asp, accessed 15 April 2007.

5 Center for Engineering and Science, http://onlineethics.org/glossary.html#anchT, accessed 6 February 2007.

6 Ibid. It is, of course, possible to rely on other people or on circumstances simply because one lacks other options.

7 The unit received a warning in mid-September 1993 and was sufficiently deployed by 15 November to start operations.

8 I wanted to get the C&P through the approval process quickly because the unit's change of command ceremony was coming up and it was likely that the new CO would not be keen on disciplining a soldier the day after he took over command. The unit was slated to go on block leave for three weeks, so if it were not done before that, it would be almost a month before we could get everything lined up again. As a result, I felt it would be better to get the outgoing CO, who at any rate would be more familiar with the case, to sign off on the C&P.

9 At the time, NDHQ did not keep track of recorded warnings.

10 During the interviews, some soldiers expressed great relief that someone was now looking into the problems within the company. All those that gave information expressed concern that if word got out that they had said something, the miscreant group would target them.

11 Although I did not realize the significance of the revelations regarding discipline at the time, it was an important discovery. Regardless of the reasons for such a soft attitude towards discipline, the effect was to take away from those leaders that wanted to maintain discipline the necessary authority and tools to do so. The lack of enforced discipline likely allowed the core group of five, who were little better than criminals, to fill this vacuum by establishing an informal chain of command where they were free to do what they wanted without fear of being punished.

12 As the meeting ended and we were moving towards the door, the CO asked me how far up the chain of command I thought this thing went. I paused for a moment, finally stating that I was not sure. I went on to say that the 2IC was new and that Lieutenant Walden and the CSM had brought the issue to light, so I knew that key members at the top were on side. I also said that it was encouraging to note that with the exception of one NCM, the interviews did not bring to light any specific issues with individuals in leadership

positions. I then stated that as long as the key leaders within the company were on side and I had the unit's full support, then those that were part of the problem would quickly fall back into line or attempt to leave. Departing the meeting and heading back to company lines, I realized that the CO had asked an interesting and difficult question. I had developed as a leader always thinking that it was my responsibility to earn the trust of my subordinates; I had never thought that I would be in a situation where they would need to earn my trust. Unfortunately, this is exactly what had to be done in the coming weeks. With the exception of a small group that had been in on the initial unravelling of the situation, I would have to be careful about what I said and how I acted until I was sure that all subordinates were prepared to support the forthcoming changes.

13 Canada, *Leadership in the Canadian Forces: Doctrine* (Ottawa: DND, 2006), Chapter 1, online version, http://www.cda.forces.gc.ca/CFLI/engraph/leadership/doctrine/ch1_e.asp, accessed 30 January 2007. The manual goes on in detail to state, "Hence effective CF leaders at all levels: exercise good judgment in decisions that affect others and do not expose people to unnecessary risks; show trust and confidence in their subordinates by giving them additional authority and involving them in decisions where circumstances allow; demonstrate concern for the well-being of their subordinates, represent their interests, and ensure they are supported and taken care of by the organization; show consideration and respect for others, treating subordinates fairly – without favour or discrimination; focus on the mission, maintaining high standards and honest and open communications; lead by example, sharing risks and hardships and refusing to accept or take special privileges; keep their word and can be counted on to honour their obligations; and demonstrate high levels of proficiency in the performance of core functions and take advantage of opportunities to enhance their professional expertise and competence."

14 Ibid.

15 At the start of the competition, a set number of soldiers would be picked from each section or platoon and individual scores of the soldiers in that sub-unit would be combined to get an overall score.

CONTRIBUTORS

Major **Tony Balasevicius** is an infantry officer and a member of The Royal Canadian Regiment. He has held various platoon and company command positions in all three battalions of The RCR, in addition to the Canadian Airborne Regiment. He is currently a member of the directing staff with the Department of Applied Military Science at the Royal Military College of Canada in Kingston, Ontario.

Major **Brent Beardsley** has served for nearly 30 years as an infantry officer in The RCR. His service includes four tours of regimental duty in Canada, Europe and the Middle East, and a number of operational tours in Cyprus, UN Headquarters in New York and Rwanda. In 1993-1994, he served as then-Major-General Roméo Dallaire's personal staff officer in UNAMIR before and during the genocide in Rwanda and is the co-author, with Dallaire, now – Lieutenant-General (Ret'd) and senator, of the best-seller *Shake Hands with the Devil: The Failure of Humanity in Rwanda.* He is currently a research officer with the Canadian Forces Leadership Institute in Kingston. He holds a number of academic degrees that include a BA in history from Concordia University, a post-graduate diploma in education from McGill University and a master's in applied science in management from RMC. He is currently completing his MA in war studies, also through RMC, where his focus is genocide and humanitarian intervention.

Master Warrant Officer **Martin Colbert** is currently the Cadet Wing Sergeant-Major at the Royal Military College of Canada. During his 23 years in the Canadian Forces, he has served with the 5ᵉ Régiment d'Artillerie Légère du Canada, the Canadian Airborne Regiment, the Canadian Airborne Centre, and the 3ᵉ Bataillon, Royal 22ᵉ Régiment. Aside from possessing a number of airborne and jungle warfare qualifications, he has extensive operational experience, including deployments to Cyprus, Haiti, East Timor, Bosnia and Afghanistan.

Lieutenant-Colonel **Ian Hope** is currently a student at the US Army War College in Carlisle, Pennsylvania. He has served as a liaison officer to United States Central Command in Tampa, Florida and commanded Task Force Orion in Kandahar province, Afghanistan from January to August 2006 under Operation Enduring Freedom and later under ISAF IX. Lieutenant-Colonel Hope served for 17 years in infantry battalions that included the Canadian Airborne Regiment and the British Parachute

Regiment. He holds a BA (Honours) in history from Acadia University, a master's of military arts and science from the United States Army Command and General Staff College, and is a graduate of the United States Army School of Advanced Military Studies.

Colonel, Dr. **Bernd Horn** is currently the deputy commander of CANSOFCOM. He is an experienced infantry officer with command experience at the unit and sub-unit level. He was the commanding officer of 1 RCR (2001-2003); the officer commanding 3 Commando, the Canadian Airborne Regiment (1993-1995); and, the officer commanding "B" Company, 1 RCR (1992-1993). He recently served as the director of CFLI (2004-2007). Colonel Horn holds a MA and PhD in war studies from RMC, where he is an adjunct-associate professor of history, and has authored, co-authored, edited and co-edited 20 books and numerous articles on military affairs and military history.

Lieutenant-Colonel **Jim Kile** is currently the commanding officer of 2 Field Ambulance in Petawawa. A medical officer, he received his MD from the University of Toronto; he also holds a BSc in kinesiology and a MSc in work physiology. In 1995, within a year of qualifying as a physician, Lieutenant-Colonel Kile was deployed on Operation Mandarin to serve as the battalion medical officer for the Canadian Logistics Battalion as part of UNPROFOR in the Balkans. He has held a number of command and staff appointments both in Canada and overseas, including serving as the commanding officer of the field hospital in Kabul, Afghanistan. He remains clinically engaged and has attained special competency in emergency medicine.

Craig Leslie Mantle graduated from Queen's University in 2002 with his MA in Canadian military history and has been employed by CFLI ever since. He is a doctoral candidate at the University of Calgary where he is studying officer-man relations in the Canadian Expeditionary Force during the First World War. He has published a number of volumes related to disobedience, leadership at the levels of command and the participation of Aboriginal peoples in the military.

Padre **Steve Moore** is a native of Fredericton, New Brunswick. He is an ordained elder with the United Church of Canada and is presently serving as the senior chaplain of RMC. He has deployed twice with the UN, first with UNPROFOR to the former Yugoslavia (1992-1993) and then to Haiti (1997). Padre Moore received his PhD in theology from Saint-Paul's University, Ottawa. During the summer of 2006, he deployed to

Kandahar, Afghanistan for research purposes where he studied the Provincial Reconstruction Team's concept of nation building. The focus of his thesis is the role of operational chaplains and their capacity to aid conflicting ethno-religious communities in the reconciliation process via the building of relation among estranged local religious leaders, and by extension, their faith communities. He has published a number of scholarly works in *Ecumenism* and through CDA Press.

Master Corporal (Ret'd) **Shawn Raines** began his service in January 1986. After successful trades training, he was posted to 2 Service Battalion at Petawawa. From there, he served with 4 Air Defence Regiment in Lahr, Germany. While in Germany, he spent time with 4 Service Battalion, 3 RCR, and was deployed to the former Yugoslavia on rotation zero of UNPROFOR. After his return to Canada, Master Corporal Raines was posted to 8 Aircraft Maintenance Squadron at Trenton, Ontario. He also served with Land Force Central Area Headquarters/Area Support Unit in Toronto, Canadian Forces Recruiting (Toronto), and Canadian Forces Recruiting Group Headquarters at CFB Borden. His final posting was as the chief clerk of the Regional Cadet Support Unit Headquarters, again at Borden. He retired from the CF in June 2006 and is now employed as the administration officer in the Foodborne, Waterborne and Zoonotic Infection Division of the Centre for Infectious Disease Prevention and Control of the Public Health Agency of Canada.

Lieutenant-Colonel **Jeff Stouffer** enrolled in the CF in 1988 as a personnel selection officer following completion of a master's degree in psychology from the University of Manitoba. He has since served in a variety of staff and research positions, including a one-year secondment with the Royal Canadian Mounted Police. He is currently the section head of research at CFLI.

GLOSSARY
OF ABBREVIATIONS AND ACRONYMS

2IC	Second-in-Command
9/11	11 September 2001
ANSF	Afghan National Security Forces
AOR	Area of Responsibility
ATA	Afghanistan Transitional Authority
BG	Battle Group
BH	Bosnia-Herzegovina
C&P	Counselling and Probation
CANBAT II	Canadian Battalion II
CDA	Canadian Defence Academy
CF	Canadian Forces
CFLI	Canadian Forces Leadership Institute
CISS	Canadian Institute of Strategic Studies
CO	Commanding Officer
CQ	Company Quartermaster
CSIS	Center for Strategic and International Studies
CSM	Company Sergeant-Major
DCIEM	Defence and Civil Institute of Environmental Medicine
DCO	Deputy Commanding Officer
DND	Department of National Defence
DRDC	Defence Research and Development Canada
EW	Electronic Warfare
Fd Amb	Field Ambulance
GT	Groupement Tactique
HQ	Headquarters
HSS Coy	Health Services Support Company
HUMINT	Human Intelligence

ICOM	Communication Intercept
IED	Improvised Explosive Device
IFF	Identification Friend or Foe
ISAF	International Security Assistance Force
ISTAR	Intelligence, Surveillance, Target Acquisition, Reconnaissance
JNA	Yugoslavian National Army
JTF-2	Joint Task Force-2
LAV	Light Armoured Vehicle
MIR	Medical Inspection Room
MP	Military Police
MWO	Master Warrant Officer
NATO	North Atlantic Treaty Organization
NCM	Non-Commissioned Member
NDHQ	National Defence Headquarters
NGO	Non-Governmental Organization
OC	Officer Commanding
Op	Operation
OR	Orderly Room
PCC	Provincial Coordination Centre
PDC	Provincial Development Committee
PPCLI	Princess Patricia's Canadian Light Infantry
PPE	Personal Protective Equipment
PRT	Provincial Reconstruction Team
R22ndR	Royal 22nd Regiment
RCR	The Royal Canadian Regiment
Ret'd	Retired
ROE	Rules of Engagement
Roto	Rotation
RSM	Regimental Sergeant-Major
RW	Recorded Warning

SQFT	Secteur du Québec de la Force terrestre
Svc Bn	Service Battalion
TF	Task Force
TFK	Task Force Kabul
TFO	Task Force Orion
UN	United Nations
UNAMIR	United Nations Assistance Mission for Rwanda
UNPROFOR	United Nations Protection Force
UNSMIH	United Nations Support Mission in Haiti
US	United States
WO	Warrant Officer

INDEX